Towards a Collaborative Environment Research Agenda

Challenges for Business and Society

Edited by

Alyson Warhurst

Director, International Centre for the Environment, and
Director, Mining and Environment Research Network
University of Bath

First published in Great Britain 2000 by
MACMILLAN PRESS LTD
Houndmills, Basingstoke, Hampshire RG21 6XS and London
Companies and representatives throughout the world

A catalogue record for this book is available from the British Library.

ISBN 0–333–67479–0

First published in the United States of America 2000 by
ST. MARTIN'S PRESS, INC.,
Scholarly and Reference Division,
175 Fifth Avenue, New York, N.Y. 10010

ISBN 0–312–23011–7

Library of Congress Cataloging-in-Publication Data
Towards a collaborative environment research agenda : challenges for
business and society / edited by Alyson Warhurst.
p. cm.
Includes bibliographical references and index.
ISBN 0–312–23011–7 (cloth)
1. Environmental policy. 2. Pollution. 3. Environmental management.
I. Warhurst, Alyson.

GE170 .T68 1999
363.7'05—dc21   99–048183

This book is printed on paper suitable for recycling and made from fully managed and
sustained forest sources.

10   9   8   7   6   5   4   3   2   1
09   08   07   06   05   04   03   02   01   00

Printed and bound in Great Britain by
Antony Rowe Ltd, Chippenham, Wiltshire

Contents

Acknowledgements

I would like to acknowledge the excellent and painstaking editorial assistance provided by Claire Brignall in the preparation of this text. Claire's editorial skills from her national press days combined with her environmental expertise contributed to the successful implementation of this ICE project.

ALYSON WARHURST

The editor, contributors and publishers are grateful to the following for permission to reproduce copyright material: International Development Research Centre of Canada for material in Chapter 1. Every effort has been made to contact all the copyright-holders—if any have been inadvertently omitted the publishers will be pleased to make the necessary arrangement at the earliest opportunity.

Notes on the Contributors

William Bains held a lectureship at the University of Bath, researching *inter alia* genomics and bioinformatics technologies. In 1988 he joined a technology consultancy to do much the same thing for industry, and to provide evaluations of how the maturing molecular technologies could be used to build businesses. In October 1996, he joined Merlin Ventures to help discover science ready to be developed in new start-up companies, and to run Merlin's scientific team. He was awarded the Toshiba Invention Award in 1992 for his development of environmental sensing systems, and was elected to the Human Genome Organization (HUGO) in 1994.

David Blakesley is currently a senior researcher at Horticulture Research International, East Malling. Until 1997 he was a member of academic staff in the School of Biology and Biochemistry at the University of Bath. His research interests include tree propagation and the restoration of natural forest ecosystems, and biotechnology for the conservation and improvement of tree and crop genetic resources. He has worked closely with a forest restoration research programme in Thailand since its inception in 1992, and has made seven visits to the project. He is currently working with Shell International Renewables on a genetic improvement programme for forestry and biomass tree species.

Ingolfur Blühdorn is a Lecturer in Politics in the Department of European Studies and Modern Languages at the University of Bath. He has an MA from the University of Nuremberg (Germany). His main area of research is environmental sociology. In recent years he has published a number of articles on European environmental movements and the changing perception of environmental problems. He is currently working on a major book on changes in European environmentalism which he describes as a shift towards 'post-ecologist' politics.

Gwendolyn Brandon is a research officer currently involved in an EPSRC project on Smart Meters and their ramifications for the promotion of more sustainable domestic energy consumption, which extends her general research interests on the links between lifestyles and sustainability. She has a BSc in Rural Environment Studies from London University (Wye College) and a PhD from Imperial College (Centre for Environmental Technology) examining public environmental concerns and related behaviour.

Alan Day is Head of the Department of Architecture and Civil Engineering at the University of Bath. He studied architecture in Glasgow and at the Architectural Association in London. His research interests are concerned with the application of computers to architecture and urban design and he has been responsible for the construction of a detailed three-dimensional computer model of the City of Bath. For the past three years Alan has been carrying out research funded EPSCR under their Sustainable Cities programme examining the impact of feedback on domestic energy consumption.

Steven Elliott is a Lecturer in Ecology in the Department of Biology, Faculty of Science, Chiang Mai University, Thailand. He is currently a Co-director of the Forest Restoration Research Unit in Chiang Mai, Thailand. He previously worked on medicinal plants in Gunung Leuser National Park, Indonesia, and the ecology of the western lowland gorilla, Gabon, before joining Chiang Mai University in 1986. His research interests include the restoration of natural forest ecosystems, and he has published widely in this area. He is also very interested in education, and the transfer of tree propagation/forest restoration technology, and has worked closely with local villagers, schools and other local organisations.

Geoffrey P. Hammond is Professor of Mechanical Engineering and Head of the Thermofluids Group at the University of Bath. He is a mechanical engineer with a multidisciplinary background, including environmental engineering and management. During the 1960s and early 1970s he worked as a design and development engineer in the UK refrigeration industry, before commencing an academic career at Uganda Technical College (under the auspices of Voluntary Service Overseas) teaching mainly in the field of applied thermodynamics. He held various academic appointments within the Applied Energy Group at Cranfield University (1976–1989) before moving to the University of Bath, where he took up a new Professorship partially supported by British Gas plc. Geoffrey Hammond's research interests are mainly in the areas of energy, environment, and sustainable development, and thermo-fluids and heat transfer. He is the author of a number of research papers in these areas, and was the joint recipient of the Dufton silver medal for a paper on heat transfer that was published in one of the Engineering Institutions' research journals. Professor Hammond has been a member of many Research Council committees covering the built environment and computational modelling. Outside the University he is a Patron and Trustee of the Bath Environment Centre Limited.

Christopher J. Heady is Professor of Applied Economics at the University of Bath and Research Associate at the Institute for Fiscal Studies. He is referee for numerous journals including: *Bulletin of Economic Research*; *Canadian Journal of Economics*; *Economica*; *Fiscal Studies*; *Journal of International Economics*, and the *World Bank Economic Review*. Professor Heady graduated from Cambridge and was awarded the Wrenbury Scholarship in Political Economy. He entered Yale University Graduate School with a Fellowship and gained his MA, MPhil and PhD in economics. He has lectured widely, both in the UK and overseas and has also worked extensively in the private sector carrying out a variety of consultancy projects. In 1994, Professor Heady prepared a paper on 'The financing of labor market insurance' for the Education and Social Policy Development Department at The World Bank. More recently he has assisted the Asian Development Bank as a Mission Member undertaking a study of sub-provincial fiscal relations in China.

Alan Lewis is Director of the Bath University Centre for Economic Psychology, Professor in the Psychology department and editor in chief of the *Journal of Economic Psychology*. His major publications include *Ethics and Economic Affairs* (ed. with Karl-Erik Warneryd, 1994); *The New Economic Mind* (with Paul Webley and Adrian Furnham 1995); and (with Craig Mackenzie, Paul Webley and Adrian Winnett) 'Morals and markets: some policy implications of ethical investing' in Peter Taylor-Gooby (ed.) *Choice and Public Policy*, 1998.

J. Allister McGregor is a Lecturer in Development Administration and is based in the Department of Economics and International Development at the University of Bath. He is currently co-ordinator of a joint project, with the Institute of Development Studies in Nairobi, Kenya, on the monitoring and evaluation of small-enterprise development. He was also awarded a Research Scholarship in Rural Sociology at the Agricultural University, Wageningen in the Netherlands. Allister is currently researching the following: the development of Southern Thailand and the implications of this for the social and economic organization of rural communities; the role of rural financial markets in developing countries; policy issues and the economics of private and public banks in development; and the political economy of agrarian change in Bangladesh.

Stuart A. MacGregor is a lecturer based in the Department of Mechanical Engineering at the University of Bath. He is a member of both the Institution of Mechanical Engineers and the Institute of Energy. Stuart has a BSc Honours in Mechanical Engineering. He also has a PhD entitled, 'Performance and characteristics of axial vortex amplifiers' from University College Cardiff. Dr MacGregor's research interests and experience lie mainly in the field of experimental fluid mechanics, especially laser and thermal anemometry. While at Bath he has applied these techniques to a number of projects which include the development of turbocharger turbines and compressors, duct flows, the internal flow structure of indirect injection diesel engines and convective heat transfer from electronic components. He also has interests in computational fluid dynamics (CFD), especially applied to IC engines. Recent projects in this area include indirect injection diesel engines, 2-stroke scavenging, the flows in intercooler ducts and the integration of CFD with CAD applied to direct injection diesel engine port design.

Malcolm McIntosh has been the Director of the Corporate Citizenship Unit at Warwick Business School since 1999. He is the co-author of *Corporate Citizenship: Successful Strategies for Responsible Companies*, published by FT Management in 1998. He is also the editor of *Visions of Ethical Business*, distributed free by FT Management and PriceWaterhouseCoopers containing 'thinkpieces' on the future of socially responsible business by leading world figures.

Chris Oulton is Head of Initial Teacher Education at University College Worcester. Until October 1998 he was a member of the Centre for Research in Environmental Education Theory and Practice at the University of Bath. His research interests are environmental education in the pre-service training of teachers and values education and environmental education. Working with Bill Scott,

Chris has published widely in these areas. From 1993 to 1996 he was Chair of the Association for Teacher Education in Europe (ATEE) Working Group on Environmental Education and Teacher Education. Chris is co-editor of the journal Environmental Education Research.

Alan D. M. Rayner has been a Reader in the Department of Biology and Biochemistry at the University of Bath since 1985. He has held various honours including, BP Venture Research Fellow 1987–1991, Visiting Miller Research Professor, University of California at Berkeley, 1994, and Centenary Fellow of the British Mycological Society, 1996. He is currently President of the British Mycological Society. He has published over 120 research papers and articles as well as six books. Dr Rayner's interests range from subcellular to ecosystem-level scales of organization. He is also a keen field biologist with a good knowledge of plant animal and fungal ecology, diversity and classification. He has used his wide interests in the past to energise some fundamental aspects of fungal biology that have been the main focus of his research, using methodologies, helped by various collaborations, ranging from mathematical to molecular.

William Scott is Head of the Department for Education at the University of Bath, where he has worked since 1978. During this time he has been Course Director for both undergraduate and postgraduate teaching programmes and has conducted a range of externally funded evaluation studies on behalf of government and other agencies. He is currently Director of Studies for the Department's Doctorate in Education Research programme: the EdD. He is a member of the Culture and Environmental research group within the Department of Education, and Director of the Centre for Research in Environmental Education Theory and Practice (CREE). He co-edits *Environmental Education Research*, edits *Assessment and Evaluation in Higher Education*, is a Fellow of the Royal Society of Arts, and a member of the North American Commission on Environmental Education Research.

Joseph Szarka is Senior Lecturer in French Studies at the University of Bath. Having worked in industry, he moved into lecturing and has taught and researched at the Cambridgeshire College of Arts and Technology, Newcastle upon Tyne Polytechnic and Aston University. He joined the University of Bath in 1987. He has written extensively on various aspects of political economy, electoral politics and environmental policy-making. Major publications include *Business in France* (1992) and *Entrepreneurial Textile Communities*—in collaboration with M. Pitt and A. Bull (1993). He is currently completing a book entitled *The Shaping of Environmental Policy in France.*

Alyson Warhurst is Director of the International Centre for the Environment (ICE) and the Mining and Environment Network (MERN) at the University of Bath. She has an established academic reputation in undertaking research at the public policy/corporate strategy interface, and has lectured and published widely in this field. Alyson has worked extensively in Latin America and China, and gained valuable research management and networking experience while working for the International Development Research Centre (Canadian Aid Programme). Her expertise has bridged both the earth and social sciences resulting in the development of an interdisciplinary understanding of the 'Environmental Imperative' as it affects key industrial sectors and social groups in both the industrialised and developing world. A specific contribution to the field of minerals policy and business has been to conduct and direct research that analyses the relationship between environmental regulation, technological change, competitiveness and environmental practices at the firm level, in different socio-economic contexts.

Adrian Winnett is a Lecturer in Economics in the Department of Economics and International Development and a member of the Centre for Development Studies and the Centre for Economic Psychology at the University of Bath. In recent years he has also worked in Bangladesh, Indonesia, and, especially, Thailand. Adrian's main areas of research are in the management of natural resources, and more broadly in problems of sustainable development, especially in South and South East Asia. He has taught extensively on environmental economics at both undergraduate and postgraduate level, in both the UK and Thailand.

Introduction: Towards a Collaborative Environment Research Agenda

Alyson Warhurst

I THE INTERNATIONAL CENTRE FOR THE ENVIRONMENT

The chapters in this volume are based on presentations made within a seminar series initiated by the International Centre for the Environment (ICE). ICE was established in 1995 and is an inter-departmental virtual centre located in the School of Management at the University of Bath. The goal of the Centre is to promote and expand the environmental research and education capacity of the University. The Centre works as a network of 'affiliated' staff from departments across the University to provide a dynamic intellectual focus, which has the strong back-up of a wide portfolio of achievement in environmental research and education that is characterized by substantial interdisciplinarity. It has achieved a critical mass of expertise in this way rather than through seeking to establish a separate environment department. As a result of this policy decision, ICE remains one of the few environment research centres in the UK that can truly claim to be:

- multi-disciplinary based on the capacity to bring together relevant specialist expertise on a project-by-project basis
- broad-ranging with a critical mass of competence (i.e. 70 researchers)
- all-encompassing with regard to the themes that we are able to address

The combined expertise that ICE and the University are able to provide in addressing the principal environmental challenges facing society include:

1

- pollution prevention
- waste minimization
- biodiversity issues and conservation
- ecosystem management
- contaminated land and water
- ecotoxicology
- analytical methods
- sustainable development—rural and urban
- environmental performance
- regulatory effectiveness and firm strategy
- education and public perception
- ethics and corporate social responsibility
- mobility and transportation
- energy generation and conservation

II TOWARDS AN ENVIRONMENT RESEARCH AGENDA

A central activity in achieving this goal is the development and definition of a new environmental research agenda which encompasses multi-disciplinary approaches to multi-faceted environmental issues. The interdisciplinary seminar series 'Towards an Environment Research Agenda' (TERA) has successfully evolved since 1995. It has consistently been very well attended and has brought together staff and students from a number of different departments including Management, Social Sciences, Education, Chemical Engineering, Mechanical Engineering, Chemistry, Biology and Biochemistry, Materials Science, and Architecture and Civil Engineering, Pharmacy and Pharmacology and Postgraduate Medicine. Our seminar series builds on our strengths throughout the disciplines and aims to disseminate research findings, draw out the interdisciplinary links and formulate a new research agenda which is both innovative and collaborative. To assist this process, seminars are chaired and commented on by researchers or practitioners from different disciplines to the speaker who, none the less, are working on related areas of research enquiry. In recent months, the programme of external speakers has expanded considerably serving to enhance our ability to generate concrete research projects with broader lines into the local and international community.

The first series of TERA seminars had two aims. First, to introduce to a broad constituency among University staff and students and local community, the key environmental themes that University researchers are working on; and secondly, by exposing research ideas and findings to interdisciplinary enquiry and challenge, to develop an agenda for future collaborative multidisciplinary research that propels our knowledge boundaries forward. For this reason each chapter, although representing a distinct area of environmental research, draws out interdisciplinary research themes and suggests additional texts for those readers wishing to explore the subject further. This approach reflects the belief within ICE that good innovative environmental research needs to combine both disciplinary excellence—whether based in management studies, biology, economics, education or engineering—with multidisciplinary insight and context. This, we would argue, becomes more relevant the more applied, and solutions-oriented, the research.

A common theme in this first collection of chapters is therefore policy research across the three interrelated themes where the University has established a reputation for research excellence, namely:

- corporate strategy and public policy towards the environment
- clean technology and integrated environmental management
- natural resources and their management

III CORPORATE STRATEGY AND PUBLIC POLICY TOWARDS THE ENVIRONMENT

In the area of public policy and corporate strategy, Alyson Warhurst in Chapter 1, 'Environmental Regulation, Technical Change and Competitiveness within a Sustainable Economy', opens the debate with an analysis of the respective roles of environmental regulation and corporate environmental strategy in promoting sustainable development. Based on case-study research in the minerals sector she analyses the paradigm shift that has led to the development of the 'win–win' scenario concept of 'pollution prevention pays' and the new policy incentives that might best promote anticipative and preventative approaches to environmentally proficient production, as opposed to 'pollute pays', end-of-pipe and costly clean up.

Malcolm McIntosh in Chapter 2, 'Globalization and Social Responsibility: Issues in Corporate Citizenship', extends the concept of 'win–win' scenarios into the realm of corporate social responsibility and local community. He describes how different international companies, in a range of socioeconomic contexts, have responded to the new ethical imperative of social responsibility by combining profit with shared values, and by integrating business with community.

Gwendolyn Brandon and Alan Day in Chapter 3, entitled 'Making Domestic Energy Use Visible', transfer the policy debate to the local level with a strong focus on 'solutions' research. Based on an in-depth analysis of energy use in Bath households, they report on research experiments to enhance energy conservation through the installation of accessible energy consumption monitoring devices and 'smart' meters which provide energy advice as well as consumption information.

In Chapter 4, 'Green and Ethical Investing: Can It Make a difference?', Alan Lewis and Craig Mackenzie analyse the emerging trend of green and ethical investment funds. They find that some people are prepared to sacrifice a degree of financial gain in order to transform environmental beliefs into positive action and that there exists scope to widen that community through tax and other fiscal incentives. Among the policy recommendations in their research, they suggest that to be truly effective, investment funds need to actively promote improved corporate environmental performance through finance conditionality and other methods of active engagement.

Chapter 5, 'Environmental Education and Teacher Education: A Critical Review of Effective Theory and Practice', by William Scott and Christopher Oulton, makes a strong argument for the development of a research agenda to support environmental education in schools and universities, particularly with respect to teacher education programmes. They further analyse what constitutes the most appropriate environmental education strategy in particular contexts. They conclude that positive interaction between researcher and practitioner is paramount, and emphasize the importance of partnership between research in universities and teacher practice in schools.

Finally, in concluding the section on public policy and corporate strategy, Joseph Szarka in Chapter 6 analyses 'The Shaping of French Environmental Policy in the 1990s'. He describes social and economic drivers of regulatory innovation, and the adoption of European environmental directives, especially those based on market incentives. He links this to a political climate in France over the last decade that favoured the development of ecology parties, and entrepreneurial Ministers of Environment, underlined by institutional reorganization and enhanced financing.

IV CLEAN TECHNOLOGY AND INTEGRATED ENVIRONMENTAL MANAGEMENT

The ICE theme of Clean Technology and Integrated Environmental Management is introduced by Geoffrey P. Hammond, in Chapter 7 entitled 'Energy and the Environment'. He argues for an interdisciplinary approach, building on the study of thermodynamics and life-cycle assessment, and prescriptive techniques such as exergy and energy analysis. He discusses problems of energy use and demand alongside the availability of indigenous fossil fuels, nuclear power and the potential role of 'renewable' energy sources, and concludes by making recommendations to improve energy and resource use efficiency in the context of the UK government's evolving energy, environmental and transport policies.

William Bains, in Chapter 8, introduces another field of Integrated Environmental Management with regard to 'Finding Poisons: Techniques for Gross Pollution Monitoring'. He describes the benefit of biologically based sensors (e.g. trout, microbes, etc.) over chemical sensors in detecting different types of toxins. Finally he analyses some of the constraints facing industry in investing in the further development of such techniques and the need for further research and work on commercial application.

Stuart A. MacGregor, in Chapter 9, entitled 'Controlling Emissions', investigates the scope of cleaner technology to control vehicle emissions. He considers recent advances and future developments in internal combustion engine technology and suggests that interdisciplinary research, e.g. into alternative fuels and sources of energy, involving the participation of engineers and pure scientists such as mathematicians, will contribute best to the radical innovation required to control emissions effectively from motor vehicles.

V NATURAL RESOURCES AND THEIR MANAGEMENT

The ICE theme of natural resources is addressed first by Alan D. M. Rayner in Chapter 10, entitled 'Challenging Environmental Uncertainty: Dynamic Boundaries Beyond the Selfish Gene', which is an appraisal of environmental changes instigated by the complex interactions of living systems. Working from a case study of fungi responses to oxidative stress, he argues that current analytical models of evolutionary biology emphasize individual competitiveness and discreteness, rather than dynamic interaction and systems.

Ingolfur Blühdorn, in Chapter 11: 'Construction and Deconstruction: Ecological Politics after the End of Nature', then addresses the concern of sociologists to explain different societal responses to different environmental issues. He explains that for some sociologists, nature and associated ecological problems are social constructions, rather than realities external to social discourse. He therefore argues for interdisciplinary research to investigate the different extents to which the environmental agenda is shaped by both changing physical conditions and social constructions.

Chapters 12 and 13 address the broader issues of the management of natural resources in the developing country context.

Chapter 12, by David Blakesley, J. Allister McGregor and Steven Elliott, analyses the case of 'Forest Restoration Research in Protected Areas in Northern Thailand'. They describe the recent rapid loss of forest and associated biodiversity as a result of economic development and analyse solutions in terms not only of conversion to new forest but also of community programmes that involve the local population in the production of a range of forest products and ecological services. At the same time, for such an approach to be effective, Blakesley *et al.* argue for interdisciplinary research to address not only accelerated forest restoration but also education and training for local communities to ensure their full participation through improving their skills and knowledge of the issues.

Finally, Christopher Heady, J. Allister McGregor and Adrian Winnett present a comparative analysis of 'Sustainability, Access, and Equity: Tropical Floodplain Fisheries in Three Asian Countries' in Chapter 13. The research was interdisciplinary and involved biologists, economists and other social scientists using both qualitative and quantitative techniques. The chapter reports on computer simulations of different management scenarios. The authors found that sound interdisciplinary research brought into question typical assertions about over-fishing and resource depletion. The researchers found instead that there do exist sophisticated traditional controls on access to fishing and that misconstrued policies may undermine such controls on access with negative implications for income distribution particularly among vulnerable communities.

We hope you enjoy reading these contributions to environmental research and that they stimulate further interdisciplinary enquiry into the important issues they raise regarding environment and sustainable development.

Part I

Corporate Strategy and Public Policy towards the Environment

Part II

1 Environmental Regulation, Innovation and Competitiveness within a Sustainable Economy

Alyson Warhurst

Summary

Environmental policies have traditionally been guided by the 'Polluter Pays' principle, and dealt mainly with the results of environmental mismanagement—pollution—and its clean-up. The new regulatory principle of 'Pollution Prevention' aims to promote competitive and environmentally sustainable industrial production. However, as this chapter will argue, successful implementation of Pollution Prevention requires innovative research to inform the introduction of new policy mechanisms, designed both to stimulate technological innovation in firms and to encourage the commercialization and diffusion of those innovations across the boundaries of firms and nations. In order to accomplish this, policies to promote and regulate industry, which traditionally have been separate, will need to be integrated (Warhurst, 1994a). The limited ability of existing environmental regulation to encourage innovation is analysed and research priorities and policy mechanisms are suggested that may be used to stimulate the development and diffusion of clean technology, so that pollution is prevented cost-effectively from the outset, rather than cleaned up expensively once the damage is done.[1]

I INTRODUCTION: FROM LIMITS TO GROWTH TO QUALITY OF GROWTH

The 1972 Club of Rome Report, *The Limits to Growth*, predicted the imminent depletion of Earth's non-renewable resources, particularly

9

fossil fuels and metals (Meadows *et al.*, 1972). Despite such warnings, the discovery of new oil and mineral reserves, in conjunction with technical change and improved recycling, has alleviated fears of non-renewable resource depletion. As a result, the environmental debate has since shifted, and is now focused on preventing the depletion and degradation of renewable resources, such as water, air, land and bio-diversity. The growing emphasis on 'Sustainable Development' reflects this increasing concern about the degradation of renewable resources and the interaction between economic activity and environmental quality.

The 1987 Brundtland Report of the World Commission on Environment and Development defines 'Sustainable Development' as 'development that meets the needs of the present without compromising the ability of future generations to meet their own needs' (World Commission on Environment and Development, 1987, p. 43). Implicit in this definition are the ideas that environmental concerns should be incorporated into economic policies, and that economic development should promote intergenerational as well as geographical equity (Jacobs, 1991). While international commitment to the goals of Sustainable Development has been strong, research efforts to develop measurable targets and policy mechanisms for the implementation of Sustainable Development have been limited.

Environmental policies have traditionally been guided by the 'Polluter Pays' principle, and dealt mainly with the results of environmental mismanagement—pollution—and its clean-up. The new regulatory principle of 'Pollution Prevention' aims to promote competitive and environmentally sustainable industrial production. However, as this chapter will argue, successful implementation of Pollution Prevention requires the introduction of new policy mechanisms, designed to stimulate technological innovation in firms and to encourage the commercialization and diffusion of those innovations across the boundaries of firms and nations. In order to accomplish this, policies to promote and regulate industry, which traditionally have been separate, will need to be integrated (Warhurst, 1994b).

Section II of this chapter analyses the policy challenge posed by Pollution Prevention approaches to environmental management. Section III evaluates the relationship between production efficiency and innovation, developing the concept of Corporate Environmental Trajectories. Case studies of mining operations, drawing on research of the Mining and Environment Research Network (MERN), are used to illustrate these arguments.[2] The limited ability of existing environmental regulation to encourage innovation is discussed in Section IV. Section V suggests policy mechanisms that may be used to stimulate the development and diffusion of clean technology.

II THE RESEARCH AND POLICY CHALLENGE OF POLLUTION PREVENTION

Pollution Prevention represents an advance over previous policies guided by the 'Polluter Pays' principle. However, in highlighting the following two flaws in regulatory approaches to Pollution Prevention, this chapter suggests the need for a new policy approach, termed 'Environmental Innovation'. In essence, the two flaws in Pollution Prevention can be characterized as described below.

First, the research conducted by the Mining and Environment Research Network indicates that those firms which pollute most are mis-managing the environment primarily because of their inability to innovate. Environmental degradation is greatest in operations with low levels of productivity, obsolete technology, limited capital, and poor human resource management. Yet under the existing regulatory regime, many older operations, incapable of technological and managerial innovation, are being driven toward bankruptcy as a result of ever-tightening environmental controls. Particularly in the developing country context, such an outcome would not only threaten the economic goals of Sustainable Development, but would also lead to further clean-up and site rehabilitation costs. In most cases, the bulk of the environmental and social costs of these shut-downs are transferred to society. In order to prevent this, regulation must be underpinned by technology policies targeted at the least efficient firms that encourage the development of the technological and managerial capabilities required for innovation. In addition, complementary policies are needed to improve access to investment resources through such mechanisms as special lines of credit, selective R&D funding, and technology investment tax-breaks.

Secondly, although the minerals sector has traditionally been characterized by low levels of research, innovation and technological change, a number of mining companies have emerged as innovators in recent years, making significant advances in technology and process management. Innovative firms are likely to be highly efficient and competent in the management of environmental matters. These firms are able to harness both technological and organizational change to reduce the production costs as well as the environmental damage costs of their operations. Furthermore, whereas environmental compliance represents a cost that could reduce competitiveness, the dynamic firm is able to offset regulatory costs with improved production efficiency.[3]

Existing regulation often requires these innovative firms to take specific, pre-determined measures to reduce pollution at source, even if the firm is working constantly to optimize for example, metal recovery, reagent use, energy efficiency, and water conservation as part of a corporate strategy to increase competitiveness. For firms such as these, command and control requirements may stifle innovation. Regulation targeted at these dynamic companies may therefore serve its objectives better if underpinned by technology policy mechanisms and economic incentives. Such instruments would aim to: (a) reward innovation in clean technology and the adoption and diffusion of these innovations in the form of financial incentives; (b) stimulate profitable innovation in that part of the pollution cycle involving waste management, including re-mining, reagent and metals recovery, biotechnology for waste treatment, etc; (c) improve training of regulators in corporate strategy and engineering, so that appropriate regulatory responses can be crafted which will enhance the competitive advantage of dynamic firms.

The analysis presented here recognizes that mining—the case-study sector considered here—is a highly heterogeneous activity, in which source material and working environments can vary greatly both spatially and temporally. Certain emissions from minerals processing can be prevented and some 'by-products' from mining can be treated, recovered, or recycled. However, the removal and disposal of large tonnages of rock is usually inevitable in the winning of metals (Winters and Marshall, 1991; Tilton, 1994). This analysis also recognizes that radical technological and organizational innovation can change the broader context within which metals production and subsequent pollution takes place.

Environmental regulation is here to stay, and regulatory frameworks affecting mining and mineral processing activities are growing in number and complexity. Therefore, an increasing share of the metals market will belong to those companies that are playing a role in changing the industry's production parameters, and those that used their innovative capabilities to their competitive advantage. Those companies that avoided environmental control, only to be later forced to internalize the high costs of having done so, will be at a relative disadvantage.

Broadening the range of regulatory goals, and the technology policy mechanisms and economic instruments which would need to be in place to support them, as proposed here, would comprise a new policy approach towards regulating and promoting industrial development. Pollution Prevention at source would play a key role in this policy, but would not always take priority over Environmental Innovation in the competitive and environmentally sustainable development of industrial production in developing and industrialized countries.

In requiring the reduction of pollution at source, Pollution Prevention often necessitates technological and organizational changes in the production process. These changes require the development of new technological and managerial capabilities within the firm, alliances with equipment suppliers, and collaboration with R&D institutions. In turn, new policy mechanisms must encourage and reward innovation and collaboration along these lines.

The reasons for this are rooted in the determinants of environmental management in the firm. Indeed, MERN's research suggests that the environmental performance of a mining enterprise is more closely related to its innovative capacity than to the regulatory regime under which it operates (Lagos, 1992; Acero, 1993; Lin, 1992; Loayza, 1993; Warhurst, 1994a). Capacity to innovate is, in turn, related to the entrepreneurial characteristics of the firm's management, its access to capital, technological resources and skills, and the broader policy and economic environment in which it operates. This evidence suggests that technical change, stimulated by the 'Environmental Imperative', is reducing both production and environmental costs to the advantage of those dynamic companies which possess the competence and resources to innovate. Such companies include mining enterprises in developing countries, as well as transnational firms. However, the evidence is strongest for large new investment projects and greenfield sites.

In older, ongoing operations, environmental performance correlates closely with production efficiency, and environmental degradation is greatest in operations working with obsolete technology, limited capital and poor human resource management. The development of technological and managerial capabilities to effect technical change in such organizations would clearly lead to improved efficiencies in the use of energy and chemical reagents, as well as to higher metal recovery. Thus, improved production efficiency would result in improved overall environmental management, including better workplace health and safety.

At the design stage of new projects, mechanisms to achieve compliance with tightening regulatory requirements may be incorporated relatively easily. However, older, inefficient, and ongoing operations may have a much more difficult time complying with these standards. Controlling pollution problems in these cases often requires costly add-on solutions: water treatment plants, strengthening and rebuilding tailings dams, scrubbers and dust precipitators, etc. Furthermore, in the absence of technological and managerial capabilities, there is no guarantee that such items of pollution control—environmental hardware—will be incorporated into or operated effectively during the production process.

III RESEARCH, INNOVATION AND ENVIRONMENTAL PERFORMANCE

As introduced above, the evidence of empirical research undertaken by MERN suggests that many mining companies are making progress towards controlling and minimizing the environmental impacts of minerals production. Improved environmental performance at mining enterprises such as these seems to be more closely related to innovative capacity than to the regulatory regime within which the mine operates. There exists as much disparity in environmental practices between firms operating in a single regulatory regime as there is between firms operating in different regulatory regimes. Capacity to innovate is, in turn, related to the entrepreneurial characteristics of the firm's management, its access to capital, technological resources and skills, and the broader policy and economic environment in which it operates.

This same evidence suggests that the development of innovative solutions to environmental challenges is motivated by a combination of factors including public pressure, tightening regulation and increased global competition, which are collectively termed the 'Environmental Imperative'. Stimulated by the Environmental Imperative, technical change is reducing both production and environmental costs to the advantage of dynamic firms that possess the competence and resources to engage in R&D and development and innovate.

**Research Promoted Environmental Trajectories in
the Minerals Industry**

After a period of using rather 'static' technology, the mining and
mineral processing industry is currently going through a phase of technical change, as dynamic firms develop new smelting and leaching
technologies to escape economic as well as environmental constraints.
This phase of technical change has, in part, been motivated by rapidly
evolving environmental regulatory frameworks in the industrialized
countries, and is reinforced by credit conditionality in the developing
countries. In addition to these regulatory pressures, the behaviour of
a mining company also depends on a number of other factors: (a) the
mineral involved; (b) the level of integration of mining and processing activities; (c) the stage in the investment and production cycle at
which its mineral projects operate; and (d) the internal economic and
technological dynamism of the firm (i.e. whether the firm has the
financial, technical and managerial capabilities to be an innovator).
Innovation is evident particularly in the large North American and
Australian mining firms, but is increasingly becoming apparent in
developing country-based operations. However, it seems to be the
new operators and dynamic private firms which are changing their
environmental behaviour, while both state-owned enterprises and
small-scale mining groups in developing countries continue, with some
exceptions, to face constraints regarding their capacity to change environmentally damaging practices.

Dynamic firms undertaking new development projects are in the
most favourable position to invest in the R&D required to develop
more environmentally sound alternatives or to raise the capital to
purchase new technologies. Indeed, alternative mineral production
processes are being developed which have the potential to be more
economically efficient as well as environmentally less hazardous. Furthermore, firms are beginning to sell their technological innovations,
hoping to recoup their R&D costs. Motivated by the need to improve
production efficiency and environmental performance, some mining
firms have even pushed technology beyond the bounds of existing regulations. These firms often support tighter environmental regulations.
Utilizing their new environmentally sound technologies to meet
stricter standards, these firms can obtain a competitive advantage over
less technologically advanced firms in a strict regulatory environment.
Some examples of these innovative developments are given in
Section IV.

Environmental regulation is here to stay and is likely to become more widely adopted and more strictly enforced. As a result, the greater share of the metals market will not be won by those firms that avoid environmental control, only later to be forced to internalize the high cost of having done so. Instead, the competitive advantage will go to those firms that used their innovative capabilities to get ahead of the regulatory game and to push the industry's production parameters forward.

The 'Environmental Trajectories' that different mining firms might take in response to environmental and market conditions are depicted in Figure 1.1. This presents a simplified picture of different firm behaviour patterns towards the internalisation of the environmental costs resulting from their mining activities.

Figure 1.1   *Corporate environmental trajectories*

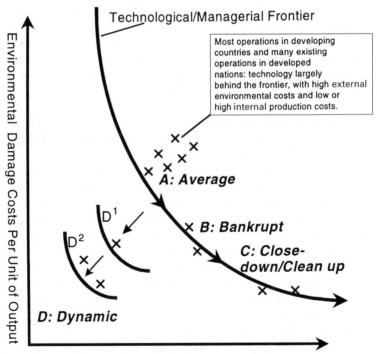

Most firms in developing countries, as well as many existing operations in developed and developing nations, are placed in a generalized group behind the technological frontier (obviously there are exceptions to this generalization) in area 'A'. Their behaviour is associated with higher environmental damages and production costs, which vary considerably depending upon the efficiency of their operations. The technological frontier is a generalized band within which most mining firms operate, to a greater or lesser extent absorbing the environmental costs resulting from their activities. Firms with ongoing operations, high sunk costs in existing facilities, and less dynamism in their technological behaviour would tend to experience environmental regulatory pressures as a cost burden. This would push them along a trajectory from A to B, as they respond to successive regulations incrementally with expensive add-on controls. Some of these firms fail to overcome these cost pressures and continue along a trajectory from bankruptcy to close-down (a trajectory toward C). Many of these firms will leave behind a legacy of environmental pollution, and the burden of clean-up will likely fall on the state and society.

Other firms are innovating by designing new technology and production practices that both lower production costs and lessen environmental damage. In doing so, they are pushing forward the technological frontier along a trajectory D_1–D_2. In addition to reaping the benefits of increased productivity, these firms are protecting themselves from having to install costly add-on technologies and undertake land rehabilitation. Greenfield operations, in particular, display high levels of dynamism, incorporating practices and technologies which yield both improved economic and environmental efficiencies.

Market and regulatory pressures suggest that the average firm will only survive in a new technological regime if it innovates. As firms advance the technological/managerial frontier regulatory requirements often advance along parallel lines. Often, dynamic firms push for strict requirements in an effort to maximize their competitive advantage over less efficient firms. As a result of advancing regulation, even previously dynamic firms must continue to innovate along a trajectory from A to D.

While tightening regulation may encourage better environmental performance, a certain amount of restraint may be warranted in the regulation of the minerals industry. As described above, regulatory burdens may lead to bankruptcy and close-down, but this does not signal the end of environmental degradation at a mine site. Pollution from abandoned sites can continue for many years, necessitating a new phase of environmental management—decommissioning, clean-up and rehabilitation. These management processes pose significant costs, which must be borne by the public sector if no one else can be held accountable. In few countries are operators liable for the clean-up of their 'sins of the past'. Therefore, unless firms are financially, technologically and managerially capable of meeting new regulatory requirements, increasing the stringency of environmental regulation has the potential to increase the overall extent of environmental degradation.

The challenge for regulators is to encourage and reward the development and implementation of innovative pollution control technologies and management practices while recognizing that increasingly stringent regulation may drive firms into bankruptcy. The ultimate goal of regulation should be to encourage operations to reduce their pollution at source, profitably clean-up pollution that 'escapes', and generate increasing amounts of economic wealth. In essence, the promotion of Environmental Innovation should be a focal point for regulators.

Environmental Research and Innovation: Case Study Analysis

As mentioned above, a growing number of dynamic innovative companies are making new investments in research leading to improved environmental management. These firms see an evolution towards stricter environmental regulation, and feel that it will be to their competitive advantage to push technological and managerial frontiers forward. Relatively free of the encumbrance of sunken investments in pollutant-producing obsolete technology and with sufficient resources for R&D and technology acquisition, they have chosen either to develop cleaner process alternatives or to purchase new improved technologies from mining equipment suppliers (who are themselves busy innovating). Increasingly, new investment projects are incorporating both improved economic and environmental efficiencies into production processes, not just in terms of new plant or equipment, but also through the development of improved management and organizational practices. Some examples of these are discussed below.

Innovation to reduce smelter emissions

The non-ferrous metals industry produces a range of metals from sulphide ores, notably nickel, copper, zinc and lead. Considerable innovations have been made over the last few years in the redesign of the production process for sulphide ores to facilitate sulphur dioxide capture and its efficient conversion. In conjunction with a steep rise in energy prices during the 1970s, the demonstration of the linkage between sulphur dioxide (SO_2) emissions and acid precipitation challenged the smelting industry to find ways of both reducing SO_2 emissions and improving energy efficiency. In seeking to meet the challenges posed by the Environmental Imperative, new technologies have been developed that improve process efficiency and cut emissions. This has been accomplished by reducing the number of stages in the smelting process, increasing the concentration of SO_2 in the off-gas, and enclosing the process so as to make the capture of off-gases as efficient as possible.

Inco Ltd—flash smelting technology (Ontario, Canada). Inco's development of oxygen flash smelting is an example of radical technical change in the face of limited possibilities for efficiency improvements in conventional technology. At one time one of the world's highest cost nickel producers, Inco was until recently the greatest single point source of environmental pollution in North America. This was due to an aged and inefficient reverberatory furnace smelter in use at Inco's Sudbury operation in Ontario, which produced excessive SO_2 emissions. The operation had reached the limits of efficiency improvements, and was still unable to meet regulatory requirements made increasingly stringent by the Ontario Environment Ministry as part of an SO_2 abatement programme to control acid rain. These factors prompted Inco to invest over CA\$3 billion in an R&D and technological innovation programme (Aitken, 1990). Under the Canadian acid rain control programme, Inco was required to reduce SO_2 emissions from the Sudbury complex by 60%. To achieve this reduction, Inco spent CA\$69 million to modernize milling and concentrating operations and CA\$425 million for smelter SO_2 abatement. The modernization process included replacement of its reverberatory furnaces with a new innovative oxygen flash smelter, a new sulphuric acid recovery plant, and an additional oxygen plant. The Inco oxygen flash smelter produces a concentrated SO_2 off-gas stream that can be efficiently captured and fixed as sulphuric acid. In addition, the flash smelting process utilizes the exothermic properties of sulphide ores and requires very little additional fuel. This process has not only reduced emissions by over 100,000 tonne/year but has also helped to transform Inco into one of the world's lowest cost nickel producers (Warhurst and Bridge, 1997).

Kennecott/Outokumpu Oy flash smelter (Garfield, Utah, USA). A flash smelter/flash converter developed by Kennecott and Outokumpu Oy at Garfield, Utah has been heralded as the 'cleanest smelter in the world'. As such, it is one of the most significant innovations in the minerals industry in recent times. The new smelter and converter complex replaces a facility which was able to handle only 60% of the concentrates produced at the Bingham Canyon mine. To meet the increasingly stringent US air quality standards, the company was faced with a choice of investing $150 million in pollution control technology for their existing smelter capacity, or investing $880 million on a new process. The new process increases the capacity of the smelter to handle 100% of the concentrates, thereby eliminating transportation and processing costs associated with the shipment of concentrates to Pacific Rim smelters. The process also enables the plant to meet or exceed all existing and anticipated air quality regulations. It is estimated that the new plant will reduce operating costs by 53% (Dimock, 1995). The principal features of the new complex are the replacement of traditional Pierce-Smith converters with a patented flash converter, the total enclosure of the converter, and the replacement of open-air transfer of molten matte with a solid-state transfer process. Molten matte is cooled with water into a granulated form prior to transfer to the converters, significantly reducing the release of sulphur dioxide and other gases in the transfer process. Although the cooling of the matte involves a loss of heat energy, 'waste' heat is captured as steam and fed to a co-generation unit. This use of flash conversion enables a continuous high throughput of material and an increased concentration of sulphur in the off-gas, greatly improving the efficiency of sulphur capture. In combination with the world's largest double contact acid de-sulpurization plant, annual average emissions of sulphur dioxide will be reduced from 3,600 pounds per hour to 200 per hour (Chiaro, 1994; Dimock, 1995; Kosich, 1995).

Pollution prevention in gold extraction

Homestake's McLaughlin gold mine (Lower Lake, California, USA). Opened in 1988, the McLaughlin gold mine is a good example of a new mine and processing facility which has been designed, constructed and operated from the outset within the bounds of the world's strictest environmental regime (Warhurst, 1992c). Although mining ceased during 1996, stockpiled ore is being processed as reclamation of the site continues. Environmental efficiency was built into every aspect of the gold mining process. Innovative process design criteria, fail-safe tailings and waste disposal systems and extensive ongoing mine rehabilitation and environmental monitoring systems characterize the McLaughlin site. The mining operation, therefore, combines a myriad of innovative technologies to define 'best practice' in environmental management. For example, before the mining operation began, an extensive environmental impact analysis was undertaken. All plant and animal species were identified and relocated, ready for rehabilitation on the completion of mining operations. The survey also measured in detail prior air, soil and water quality characteristics and water flow patterns to provide the baseline for future monitoring programmes. Assaying was undertaken not just of the gold ore, but also of the different types of gangue material, so that waste of different chemical compositions could be mined selectively and dumped in specific combinations to reduce its acid mine drainage generating capacity. Local climate conditions were evaluated to determine the frequency of water spraying needed to reduce dust, and evaporation rates were evaluated to assist water conservation and to control the water content and flood risk potential of tailings ponds. The tailings ponds themselves were constructed on specially layered impermeable natural and artificial filters, with high banking to prevent overflow and with secondary impermeable collecting ponds in the rare case of flooding.

Unlike many other mining projects where rehabilitation is seen as a costly task to be undertaken at the end of a mining operation, rehabilitation began immediately and is an ongoing activity. Not only does this serve to spread expenditure more evenly over the life of the mine but it also enables more efficient utilization of truck and earth moving capacity, as well as relevant construction personnel. When waste piles have reached a certain pre-determined dimension, soils (previously stripped from the mine area and stored) are laid down and revegetation is begun. Extensive areas of overburden and waste have already been successfully revegetated—immediately reducing environmental degradation and negative visual impacts. In addition to these build-in environmental control mechanisms, Homestake Mining Company has sophisticated environmental monitoring procedures in place. Emission irregularities, and wildlife and vegetation effects can be detected and rectified immediately, reducing the long-term risk of expensive shut-downs, costly court cases and the need for additional clean-up technologies.

Innovative waste management

In the minerals industry, it is almost impossible to avoid the production of large volumes of waste rock in the form of overburden, marginal ore dumps, tailings and slags (Gray, 1992). Often, toxicity associated with this rock waste involves the loss of expensive chemical reagents or valuable metals. This presents great opportunities for process innovation to improve efficiency and lessen environmental impact. However, there is limited public support for direct R&D into waste toxicity reduction and treatment innovations, such as the application of biotechnology to waste treatment (Warhurst, 1991b). As a result of giving such low priority to waste treatment, there is a danger that innovation in this field may not receive appropriate encouragement. This again suggests that the policy focus of Pollution Prevention needs to address the broader process of environmental innovation, rather than strictly pollution reduction at source. The following two examples demonstrate how innovation can contribute to reducing pollution through the integration of waste management into the production process. This integration does not always represent an add-on regulatory cost burden, and indeed suggests there are competitive advantages as well as environmental benefits to such a strategy.

Water treatment of Exxon's Mine (Los Bronces, Chile).   A mining project in Chile, Los Bronces, is being expanded into one of the largest open-pit copper mines in the world and consequently requires the stripping of very large tonnages of overburden and low-grade ore. Before mine development, the Chilean government warned Exxon that it would impose financial penalties for the water treatment costs on account of expected acid mine drainage from the overburden of low-grade ore dumps into the Mantaro River, the source of Santiago's drinking water. This threat became the economic justification for a bacterial leaching project at the mine. Indeed, the feasibility of this bacterial leaching project was particularly illustrative of the profitability of leaching copper from waste while preventing pollution (acid mine drainage). Over a billion tons of waste and marginal ore below the 0.6% copper cut-off grade are expected to be dumped during the project's lifetime. The waste would have an average grade of 0.25% copper and would therefore contain a lucrative 2.5 million tons of metal worth approximately US$3.5 billion, at current 1985 prices (Warhurst, 1990). The study demonstrated that with a 25% recovery, high quality cathode copper could be produced profitably, at 39 cents/lb, by re-cycling mine and dump drainage waters over a 20-year period. This was shown to have the double advantage of both extracting extra copper and avoiding government charges for water treatment. At the same time both investment and operating costs were less than two-thirds of estimated costs for a conventional water treatment plant, which would not have generated saleable copper. The Los Bronces mine thus demonstrates the potential economical benefits of building environmental controls into mine development.

In conclusion, these few examples suggest that dynamic companies are not closing down, re-investing elsewhere or exporting pollution to less restrictive regulatory regimes in developing countries. Rather they are adapting to environmental regulatory pressures by innovating, improving and commercializing their environmental technology and management practices at home and abroad.

IV   THE LIMITATIONS OF EXISTING ENVIRONMENTAL REGULATION IN PROMOTING RESEARCH AND INNOVATION

In developed and developing countries alike, environmental regulation is seen as a fundamental tool in controlling and distributing the environmental costs involved in minerals production. Existing regulation attempts to do this through a variety of mechanisms. Some regulations require the use of pre-determined technological devices and environmental management procedures to clean up and/or prevent pollution. Other regulatory mechanisms aim to force the polluter to internalize the costs of environmental pollution, through a system of fees, taxes. This second method is based on economic arguments that suggest that if forced to internalize the costs of environmental pollution, firms would reduce emissions as an element of their competitive strategy. This is described in greater detail in Box 1.1. However, there are limitations to these mechanisms, particularly in that they fail to adequately encourage and reward innovation in pollution-prevention technologies and environmental management.

Box 1.1   Environmental externalities

Macroeconomics suggests that there are two types of costs incurred in industrial production (Tilton, 1994). First, there are those costs associated with production: labour, capital and material inputs. Secondly, there are those that the producing company usually does not pay because they are external to the firm. These externalities include the costs of environmental damage such as ecological degradation, water pollution or air contamination. The following model illustrates the behaviour of a firm that is able to externalise the environmental costs of its emissions.

The Marginal Social Costs and Marginal Social Benefits of Pollution

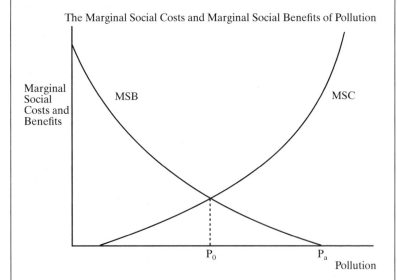

Source: Tilton (1992) 'Mining Waste, The Polluter Pays Principle and US Environmental Policy', Colorado School of Mines, Department of Mineral Economics, Working Paper 92-8, October.

According to macroeconomic theory, pollution will occur, as long as the additional benefits (in terms of goods and services derived from polluting processes) are greater than the additional costs of one more unit of pollution (in terms of public health, aesthetic value, etc.). In economic terms, pollution will occur up to the point at which the marginal social benefits (MSB) equal the marginal social costs (MSC). If all social costs and benefits of pollution are internalized by the producing firm, it will have an incentive only to pollute up to the point P_0. However, if the firm realizes all the benefits associated with pollution, but none of the costs, it has an incentive to expand its production until the additional benefits from causing a further unit of pollution are zero. Note

Continued

that, in this circumstance, pollution has reached point P_a which is far beyond the optimal point P_o. The cost burden of this environmental damage falls on society and the pollutee pays, although to a degree costs may be distributed somewhat by the State.

Particularly in the developing country context, industry is often able to externalize environmental costs through the appropriation of undervalued resources and the shifting of environmental costs onto others. This situation can eventually lead to production inefficiencies, since 'free' environmental resources may be substituted for labour, equipment, and other inputs for which the firm must pay. For example, a firm may engage in the excessive and damaging use of water resources, rather than incorporating a water treatment and recycling plant for liquid effluents. In the long run, this ability to externalize environmental costs reduces the entrepreneurial capacity of the firm and, most importantly, acts as a disincentive to innovate.

Regulatory frameworks for safeguarding the quality and availability of land, water and air degraded as a result of mining and mineral processing activities are growing in number and complexity. This has particularly been the case in the major mineral producing countries of North America and Australia as well as Japan and Europe. A common regulatory mechanism is to set maximum permissible discharge levels or minimum levels of acceptable environmental quality. These standards must be met through the use of pre-determined technologies or procedures. Best Available Technology (BAT) standards are an example of this type of 'command-and-control' mechanism. Most major environmental regulations in developing nations include some form of command-and-control or technology-forcing measures. Four issues are particularly relevant in evaluating the effectiveness of these regulatory mechanisms to reduce environmental degradation and improve environmental management practices in metals production:

(1)   It remains the case that the polluter pays only if discovered and prosecuted. In addition, remedial action tends to only occur after the pollution problem has become apparent and has caused potentially irreversible damage. This highlights the tendency of such environmental regulations to deal with the symptoms of environmental mismanagement (pollution) rather than its causes (economic constraints, technical constraints, and lack of access to technology or information about better environmental management practices). This can be serious in some instances, because once certain types of pollution have been identified,

such as acid mine drainage, it is extremely costly and sometimes technically impossible to rectify the problem and prevent its recurrence. Certain precautionary environmental controls will only work if incorporated into a project from the outset, as is the case with the use of buffer zones to protect against leaks under leach pads and tailings ponds.

(2) BAT standards may be appropriate at plant start-up, but their specified effluent and emission levels are not necessarily achievable throughout the life of the plant. Technical problems may arise and there may be variations in the quality of concentrate or smelter feed sources. It would also be erroneous for a regulatory authority to assume standards are being met simply because a pre-selected item of technology is being utilized. In the absence of the managerial capacity to monitor and maintain equipment, there is no guarantee that items of pollution control will be incorporated or operated effectively in the production process.

(3) Related to the points above, BAT standards and environmental regulations of the command and control type tend to promote incremental add-on controls to respond to evolving regulation rather than to stimulate innovation. Once large amounts of capital have been invested in a particular technology, an operation is unlikely to abandon that technology in the face of tightening regulation. This acts as a disincentive to innovate for equipment suppliers, the mining companies and metal producers. An innovative process, the development of which has required substantial R&D resources, may be superseded by a regulatory decision about what constitutes BAT for their relevant industrial activity. Rather than imposing technology from outside the firm, regulators should encourage the generation of new technologies from within the firm. In this way, the firm will develop the technology alongside the technical and managerial skills required to manage that technology. In addition, the search for profit and cost-savings is more powerful than regulatory controls in instigating technical change. It might be argued that market-based mechanisms, technology policy which is complemented by a regulatory framework, and strong corporate environmental management strategies, can better contribute to environmental performance. Furthermore, policy should be designed to facilitate the adoption of new technologies, through access to credit.

(4) In some instances, regulatory requirements are leading to shutdowns, delays, cancellations and reduced competitiveness. In

most countries (except perhaps the USA where the state has the power to hold other parties liable for environmental damages), the public sector must cover clean-up costs when operations shut-down. Particularly in developing countries, the State often has neither the resources nor the technical capacity to deal effectively with mine site remediation. The public sector will continue to struggle with this burden, unless policy is targeted at the underlying challenge of improving production efficiency and stimulating innovative capacity.

Some progress has been made to correct existing problems in the environmental regulation. There has been growing interest in the use of market-based mechanisms, whereby the a firm must pay for the environmental damage caused by its pollution. A fundamental justification for the use of market-based incentives is that these mechanisms allow companies greater freedom to choose how best to attain a given environmental standard (OECD, 1991). It has been argued that this freedom permits solutions to environmental problems which are more economically efficient. Within the minerals industry the most common market-based measure is the posting of bonds up-front (during development) for the rehabilitation of mines on closure. This is standard practice now in Canada and Malaysia. Brazil is also considering the implementation of a mercury tax, and the USA has proposed a similar tax on cyanide. Currently, however, no government has designed an explicit set of market-based incentives for industry to innovate and develop new environmental technology.

Aside from the use of market-based mechanisms, there are several further areas where policy approaches have begun to contribute to improved environmental management practices. Increasingly private, bilateral, and multilateral credit agencies are attaching conditions of environmental performance and management to loans and grants. Lenders often require both prior environmental impact assessment and the use of best practice environmental control technologies in new minerals projects. In addition, a growing number of donor agencies, in Germany, Canada, Finland and Japan, fox example, require training in environmental management at the project level. Banks and financial institutions which fund investment projects in the developing world, such as the International Finance Corporation of the World Bank and the Japanese Overseas Economic Co-operation fund demand that operations demonstrate the implementation of best-practice technological and managerial strategies (Schmidheiny and Zorraquin, 1996). For example, the Ashanti gold mine in Ghana underwent an environmental audit which specified the selection of technological and managerial approaches to improve environmental performance as a condition of receiving a loan from the International Finance Corporation. Research into recent improvements in environmental management at Ashanti suggests that the stringency of compliance may be driven more by the conditions attached to loans than by government legislation (Metals Economics Group, 1994).

Providers of credit and insurance are increasingly aware that poor environmental performance can delay a project significantly and increase exposure to liabilities. Insurance companies offering political risk insurance to Australian mining companies, for example, have started to insist on annual reviews of environmental standards in the companies' international operations (Environment Digest, 1996). A recent illustration of the heightened sensitivity among financial institutions was provided by the decision of the Overseas Private Investment Corporation to withdraw $100 million worth of risk insurance from the Grasberg mine in Irian Jaya, ostensibly on the grounds of ecological damage to forests and river systems (Kosich, 1995). Coverage was re-instated in March 1996 on the condition that the mine operators establish a trust fund to finance environmental remediation. Brazilian banks signed a Green Protocol in November 1995, giving priority to environmentally sustainable industrial development and containing a commitment to withhold financing from companies which violate Brazilian environmental legislation (Crawford, 1996). In addition to attaching conditionality to credit, some governments, particularly Canada, promote R&D activities within industry and academic institutions through joint research initiatives. For example, Canada has extensive government-funded R&D programmes to research acid mine drainage, SO_2 emissions, the toxicity of mining pollution, and to explore new clean-up solutions for these problems.

Despite these recent developments, these approaches to encouraging sound environmental management and rewarding innovation are limited in scope and breadth. As a result, there is considerable opportunity to expand these approaches, as is argued later in this chapter.

V   POLICIES TO STIMULATE ENVIRONMENTAL R&D AND THE DIFFUSION OF INNOVATIVE CAPACITY

As we have seen, research leading to innovation will be a key element in encouraging sustainable development in the minerals industry. The current trend towards liberalized economies in the developing world presents unique opportunities for the transfer of innovative environmental technologies and management skill. The policy challenge for regulators (if their ultimate aim is sustainable development) is to keep industry sufficiently dynamic so as to encourage reduction of pollution at source, profitable treatment of waste, and clean-up of pollution on closure, while also encouraging the generation of wealth throughout an operation's life. This chapter has argued that existing environmental regulation does not adequately support pollution prevention regulatory goals. Therefore, regulation must be underpinned by two further sets of policy mechanism: (1) incentives to improve production efficiency and stimulate environmental innovation; and (2) mechanisms to stimulate the diffusion of these innovations between firms and across national boundaries. A set of recommendations along these lines is presented below.

There are two types of policy mechanisms that can be used to promote environmental innovation in industry. The first set includes expenditure programmes to support R&D, environmental engineering, clean technology development and training in environmental management. The second is a set of incentives to stimulate and reward firms for environmental innovation.

Supporting Clean Technology Development

Technology policy mechanisms to support clean technology development include targeting R&D directly at selected areas of pollution prevention. This would also include co-funding of R&D projects, requiring inter-industry and industry–academic collaboration. Such programmes could be further supported by accessible systems to disseminate information about changing technological and regulatory frontiers.

A crucial point with regard to targeting support for R&D relates to how the innovation process within industry works. Too often in policy documents, innovation is conceptualized as something which builds on R&D undertaken in government or university laboratories and then applied by industry throughout its operations. Such thinking has been reflected in EPA documents regarding the agency's aims in the area of technological innovation. There is little mention of the support of R&D and engineering in the firm; innovations are considered to originate in research institutions and universities and it is suggested that policy should focus on the latter. Evidence of how innovation takes place, however, suggests that in most cases it is industry-driven, with firms complementing in-house R&D and engineering efforts by drawing upon the additional knowledge, expertise, and technology from research institutions and other firms (Rothwell, 1992; Warhurst, 1994b).

An important element of technology policy to promote source reduction innovation should be to discourage the firms from diverting resources to compliance-related R&D, as opposed to general research into process efficiency. By focusing R&D effort on process innovation and by stressing pollution prevention at source in an overall effort to improve the efficiency of the production process, both aims should be complementary.[4]

Training and Organizational Change

Encouraging environmental innovation requires a number of important changes in thinking. A range of engineering skills must be harnessed to reduce or eliminate a pollutant at source. Whereas end-of-pipe controls tend to shift pollution from one media to another, the development of environmental innovations to prevent pollution requires a multi-media approach to design. These new technologies must be designed not only to deal with water and air quality and waste but also worker health and consumer product safety. This means that multi-disciplinary training for R&D engineers working within industry should be a critical element of environmental policy.

Training is only one part of the equation. Of equal importance is the need for organizational change. Much can be learnt from the success of 'Lean Production' and related work methods developed in Japan, such as 'just-in-time' inventory control, total quality management and statistical process control. Lean Production is driven by a simple principle: the elimination of all costs incurred which do not add competitive value to a product. Secondary principles include the reduction of waste, efficient utilization of space, elimination of inventories, and the integration of quality controls throughout the production process. The implementation of Lean Production characteristically results in the reduction of managerial roles, with increased responsibility being given to engineers and workers, and a concomitant increase of multi-task activities (Womack *et al.*, 1990). Lean Production applied in the automobile industry, principally by Japanese manufacturers, has led to significant improvements (in some cases more than twofold) in productivity, quality, product development, and model range. Productivity performance data by region indicates that, on average, European and North American car plants required, 118% and 49% (respectively) more effort to undertake the same manufacturing activities as an average Japanese plant. Such advantages have translated into both cheaper and better value products, leading to the rapid growth and supremacy of Japanese firms such as Toyota, Nissan, Honda and Mitsubishi in western markets (Graves, 1991).

The implications of applying Lean Production principles to the mining industry, or of the development of radical process innovations with similar effects, would be remarkable. A combination of markedly lower investment and production costs, combined with the shortening of mine development times and mine life could have significant implications for the competitive structure of the industry, as well as reducing associated negative environmental and social effects. In making such a radical organizational change, management training will be fundamental, and engineers and miners must be introduced to new work methods. Few mining companies have taken these ideas on board. Those that have considered alternative organizational methods include CAR (Australia), Homestake's McLaughlin Mine (California), and Scuddles Mine of the Poseidon Group in Australia. For example, Scuddles has implemented an innovative multi-skilled approach to human resource development at its underground mine in Western Australia (*Mining Magazine*, 1991).

Financial Incentives for Environmental Innovation

Changes may be necessary in taxation policy in order to promote environmental innovation. The US currently gives tax incentives to industrial operations, in the form of accelerated depreciation rates for pollution control equipment, in order to support end-of-pipe pollution control (Ashford, 1991). However, investments in new production technology are not similarly treated, so that dollar-for-dollar a firm would be better off purchasing environmental technology from outside organisations, rather than building R&D capacity in the area of pollution control. More effective economic measures would provide incentives to: (a) invest in pollution prevention technology or environmental management capacity; (b) support engineering projects and training in specific areas of environmental management; (c) post bonds up-front for future pollution prevention, or reclamation on closure. Careful consideration should be given to the use of punitive taxation on reagent use or energy use requirements. Since ore deposits differ quite dramatically with respect to geology and chemistry, the effects these taxes have on both competitiveness and firm behaviour can be quite complex. Furthermore, the imposition of new taxes on ongoing operations may be perceived by operators as prejudicial and unfair. Flexible taxation provisions that allow and encourage innovative responses by industry are needed to complement strict standards and regulatory goals.

Training for regulators, including industrial experience and salary packages commensurate with corporate counterparts, will also be an important part of successfully promoting environmental innovation. Regulators must possess an intimate knowledge of the types of gains made by firms from technological change. Armed with this knowledge, regulators will be better equipped to promote technological innovation, and will be better able to respond to innovations by adapting and tightening regulation accordingly (Milliman and Prince, 1989). Innovative firms should be able to use environmental regulations to their competitive advantage. Benefits to such firms arise from the tightening of 'technology forcing' regulation so that other companies are stimulated either to invest in new technology or purchase the innovator's technology (thus enabling the innovator to recoup some of the costs of its initial investment in R&D). Regulatory authorities need to be seen to respond in this way. Moreover the rate of technological advance in pollution control is probably, at least for the informed regulator, the most useful criterion on which to judge the effectiveness of environmental policies.[5]

An important corollary to creating incentives for innovation is to encourage innovation through regulatory 'reward'. Usually this is discussed in terms of the awarding of prizes for sound environmental management such as the EPA's recent idea of an 'Environmental Leadership' programme to reward US innovators. However, the 'reward' side of the equation needs to be more sophisticated by analysing the way commercial gains to technological innovation and technology diffusing can be realised and expanded.

Finally on this issue, incentives need to be found to stimulate auxiliary firms to develop and commercialize innovative clean-up technologies including re-mining techniques. The market for such activities is vast, particularly in developing countries, and donor agencies and development assistance grants could play a key role in stimulating such investment.[6] In the US particularly, liability regulation will need to be reassessed to remove the current barriers to re-mining and treating existing mining waste.

Policies to Promote the Diffusion of Environmental Innovations

Technological and managerial capabilities are fundamental, not only for the development and utilization of innovative technologies but also for environmental management strategies to resolve pervasive inefficiencies in existing technologies. Technology transfer and technology partnership through joint venture arrangements or strategic alliances are one way to build up technological and managerial capabilities to overcome these constraints. This is particularly true in the developing country context, although such strategic alliances are emerging in all the major mineral producing countries as described earlier in this chapter.

However, there is a need to broaden policies toward technology transfer in order to encourage the effective transfer of environmental management capability. Traditionally, technology transfer is restricted to the transfer of capital goods, engineering services, and equipment designs—the physical items of the investment. Usually, experts, trained in the operation of the plant and equipment, accompany the transfer of this physical capital. As a consequence, recipients do not develop innovative capacity and become dependent upon their suppliers to make changes or improvements to successive vintages of technology. This is particularly the case in developing nations, where political and economic disparities may reinforce this situation. New forms of technology transfer in environmental management need to encourage: (a) the development of knowledge, expertise and experience required to manage technical change—of both an incremental and radical nature; and (b) the development of human resources to implement organizational changes to improve overall production efficiency, energy efficiency, and environmental management throughout the life cycle of an operation. New policy frameworks must include innovative approaches to training and skill acquisition within industrial enterprises in the areas of environmental R&D, engineering, management, on-the-job training, trouble-shooting, repair and maintenance and environmental auditing.

In global industries such as mining, international firms have played a major role in supplying managerial and engineering expertise through joint-ventures and other collaborative arrangements. However, these contributions are usually restricted to the immediate requirements of the specific investment project or item of equipment purchased. Flows of technology are often prompted by regulatory requirements. Incremental regulation yields incremental add-on technology solution. However, there is opportunity for well-structured policies to break this pattern (Warhurst and MacDonnell, 1992). Empirical research on other sectors, however, demonstrates that there exists considerable potential to increase those contributions without adversely affecting the supplier's strategic control over its 'proprietary' technology (Bell, 1990; Warhurst, 1991a, 1991b; Auty and Warhurst, 1993).

The transfer of expertise was a central goal of China's National Offshore Oil Corporation's (CNOOC) strategy, which targeted specific major oil companies and required them to transfer the capabilities to master selected areas of technology. Another interesting example is the Zimbabwe Technical Management Training Trust. It was founded by RTZ in 1982 with the aim of training South African Development Community professionals in technical management and leadership. It effectively combines academic and on-the-job training in both home and overseas operations, providing possibilities for accelerated managerial learning by being exposed to on-the-job problem-solving situations with experienced colleagues in a range of challenging technical scenarios (Warhurst, 1991b).

It would be quite feasible to build similar in-depth training programmes, concentrating on human resource development in environmental management, into many of the proposed and prospective mineral investment projects throughout the world. Investors and technology suppliers could be selected in part based upon their proven environmental management competence and their willingness to share their expertise. It cannot be over-emphasized that all technology transfer and training efforts incur a set of costs for the supplier, and these must be covered to ensure optimal results. Funds must be allocated to cover these costs or training programmes will fail to transfer technology 'mastery'. Corporate partners, the government, and donor agencies or development banks could assist in the financing of these schemes. Moreover, a government's negotiating power over the precise objectives and scope of training programmes would be greater if it contributed financially to the transfer process.

There already exists a range of commercial channels through which mine operators can purchase capital goods, engineering services and design specifications. However, the market for knowledge and expertise, including training programmes is less mature. The active development of this market will truly 'reward' innovators of environmental technology. Bilateral and multilateral agencies, development banks and government organizations can play a major role in improving this situation. Agenda 21 proposes two programmes of relevance, which should also lead to greater involvement by industry (Skea, 1993). The first programme encourages inter-firm cooperation with government support to transfer technologies that generate less waste and increase recycling. The second programme on 'responsible entrepreneurship' encourages self-regulation, environmental R&D, worldwide corporate standards, and partnership schemes to improve access to clean technology. Moreover, Agenda 21 recognizes that effective mechanisms for technology transfer require a substantial increase in the technological capabilities of recipient countries (Barnett, 1993).

In summary, existing regulatory frameworks must be underpinned by policies to promote environmental innovation and the commercialization of innovative processes in the global markets. Programmes to promote training and R&D in the area of pollution prevention will facilitate innovation in firms, and market mechanisms can be used to reward innovative firms. The transfer of innovative technologies and management techniques to less dynamic firms in industrialized and developing nations will be fundamental to achieving the goals of sustainable development. There remains great scope for the further diffusion of knowledge and technology between firms, and between firms and regulators, across firms and national boundaries. It is argued here that it is in the interests of industry, governments and international organizations, including development banks, to work at fostering the transfer of technological and managerial innovations.

VI CONCLUSIONS: TOWARDS AN ENVIRONMENTAL RESEARCH AGENDA

This chapter has sought to highlight the importance of Environmental Innovation and to suggest research targets and policy mechanisms to encourage innovation across industry but specifically in the minerals sector. The emerging regulatory principle of 'Pollution Prevention' aims to promote the competitive and environmentally sustainable development of industrial production. However, by virtue of the requirement to reduce pollution at source, Pollution Prevention demands technological and managerial changes in the production process. This requires, to differing degrees, innovation within the firm, technological alliances with equipment suppliers, and collaboration with R&D organizations. This chapter has argued that the successful implementation of Pollution Prevention will require the development of technology policy mechanisms designed both to stimulate technological innovation and best-practice in environmental management within firms, and to encourage the commercialization and diffusion of these innovations across the boundaries of firms and nations. Research is needed to inform those technology policy mechanisms.

Those firms that pollute most are mis-managing the environment primarily because of their inability to innovate and draw upon advances made through R&D. Yet under the existing regulatory regime, many older operations, incapable of technological and managerial innovation, are being driven toward bankruptcy as a result of ever-tightening environmental controls. In order to prevent this, regulation must be underpinned by technology policies targeted at the least efficient firms that encourage the development of the technological and managerial capabilities required for innovation. In addition, research is required to underpin complementary policies that are needed to improve access to investment resources through such mechanisms as special lines of credit, selective R&D funding, and technology investment tax-breaks.

A number of mining companies emerged as innovators in recent years, making significant advances in R&D, technology development and process management. These firms are able to harness both technological and organizational change to reduce the production costs as well as the environmental damage costs of their operations. Furthermore, whereas environmental compliance represents a cost that could reduce competitiveness, the dynamic firm is able to offset regulatory costs with improved production efficiency. For firms such as these, command and control requirements may stifle innovation.

Regulation targeted at these different types of firms may therefore serve its objectives better if it were underpinned by research into technology policy mechanisms and economic instruments aimed at:

- Stimulating and rewarding innovation in Pollution Prevention through technology investment tax breaks and other taxation reforms, improved access to credit, targeted R&D support and training programmes, mandatory Pollution Prevention and reclamation plans during project development, and the mandatory posting of bonds to fund reclamation and continued monitoring.
- Stimulating profitable innovation in that part of the pollution cycle involving waste management, including re-mining, reagent and

metals recovery, and the application of biotechnology to waste treatment. This would also include the removal of legislative barriers that inhibit re-mining and waste treatment.

- Facilitating and rewarding the commercialization and diffusion of Pollution Prevention technology and work practices across the boundaries of firms and nations, using mechanisms such as credit conditionality and new approaches to technology transfer. These would include inter-firm collaboration to develop the technological and managerial capabilities to innovate, in-depth training beyond the requirements of operating skills, information dissemination programmes, etc.
- Training regulators to understand the importance of innovation and to recognize opportunities where regulation could further enhance the competitive advantage of innovative firms. The concept of Corporate Environmental Trajectories can help regulators analyse the process and effects of innovation in different economic and regulatory contexts.

Leading-edge research and innovation can change the broader context within which metals production and subsequent pollution takes place. The widespread diffusion of innovation can reward dynamic firms and stimulate further innovation, as well as contribute to the furthering of best-practice in environmental management as a route towards sustainable development. Pollution Prevention policy mechanisms would be more successful if they focus on this process of innovation. Broadening the range of regulatory goals, and developing the regulatory instruments needed to achieve them, would comprise a more integrated policy approach towards both regulating and promoting industrial development. Pollution Prevention at source would play a key role, but would not always take priority in an overall policy to promote Environmental Innovation.

Such an integrated framework, underpinned by systematic multidisciplinary research, will better serve the goals of competitive and environmentally sustainable development of industrial production in developing and industrialized countries.

Acknowledgements

The author would like to acknowledge gratefully the kind assistance of Lisa Eisen and Dr Paul Mitchell in the preparation of this chapter, and to thank Kathleen Anderson, Rod Eggert and Richard Isnor for providing reference materials and feedback. Support for the empirical research reported in this chapter was provided by a grant from the John D. and Catherine T. MacArthur Foundation to the Mining and Environment Research Network. This paper draws extensively on Alyson Warhurst, 'Environmental Regulation, Innovation, and Sustainable Development', chapter 1 in *Mining and the Environment: Case Studies from the Americas*, edited by Alyson Warhurst and published by Canada's International Development Research Centre (IDRC 1998, ISBN 0-88936-828-7), which has kindly given its permission to reuse material in this article.

Notes

1. The term 'clean technology' is used in this text to refer to industrial processes that incorporate current 'best practice' in environmental management. It is not employed literally and, indeed, more accurately describes 'cleaner' technology.
2. The term mining is used here to cover all aspects of the metals production process including mine development, extraction, smelting, re-mining and waste management.
3. In the minerals sector, for example, environmental compliance costs cannot be passed readily on to consumers, because metal prices are determined at the international level in terminal auction markets. This market does not allow firms to charge a premium for minerals extracted in an environmentally conscious manner; a unit of mineral extracted in a relatively clean manner will sell for the same price as that extracted through highly damaging production processes.
4. Rothwell (1981) distinguishes between two types of innovation by firms subjected to environmental regulation. Commercial (offensive) innovation is typical of innovation that would be conducted in the absence of environmental regulation. Compliance (defensive) innovation is what is conducted specifically for the purpose of complying with environmental standards.
5. This view is reinforced by a growing number of researchers including Orr (1976), and Milliman and Prince (1989).
6. For example, over two-thirds of the current mineral reserves of Bolivia are in dumps and tailings (Warhurst and MacDonnell, 1992). Furthermore, in many developing countries, such as Peru, there are many small and medium scale dynamic firms that supply a range of inputs to the mineral sector and could, with incentives, expand their activities to the environmental arena (Nuñez-Barriga, 1993).

Further Reading

M. PORTER and VAN DER LINDE (1995) 'Green and competitive: ending the stalemate', *Harvard Business Review*, 73 (5), Sept.–Oct., pp. 120–34.

H. FOLMER, LANDIS, H. GABEL and H. OPSCHOOR (1995) *Principles of Environmental and Resource Economics: a Guide for Students and Decision-Makers*, Cheltenham: Edward Elgar.
R. HOWES, J. SKEA and B. WHELAN (1997) *Clean and Competitive?: Motivating Environmental Performance in Industry*, London: Earthscan.
A. WARHURST and G. BRIDGE, 'Economic liberalisation, innovation, and technology transfer: opportunities for cleaner production in the minerals industry', *Natural Resources Forum*, Feb. 1997.
R. WELFORD (1995) *Environmental Strategy and Sustainable Development: the Corporate Challenge for the 21st Century*. London: Routledge.

Bibliography

L. ACERO (1993) *The Case of Bauxite, Alumina and Aluminium in Brazil.* Mining and Environment Research Network, Bath, UK, MERN Working Paper Series, No. 1.
R. AITKEN (1990) Personal communication, Inco Ltd.
N. A. ASHFORD (1991) 'Legislative approaches for encouraging clean technology'. *Technology and Industrial Health*, 7(516), pp. 335–45.
R. AUTY and A. WARHURST (1993) 'Sustainable development in mineral exporting economies'. *Resources Policy*, Mar., pp. 14–29.
A. BARNETT (1993) *Technical Co-operation, Technology Transfer and Environmentally Sustainable Development: a Background Paper for the DAC Working Party on Development Assistance and the Environment*. Paris: OECD.
M. BELL (1990) *'Continuing Industrialisation, Climate Change & Internal Technology Transfer'*, a report proposed in collaboration with the Resource Policy Group, Oslo, Norway, in *SPRU report*, December, Science Policy Research Unit, University of Sussex, UK.
P. CHIARO (1994) 'Waste minimisation and pollution prevention at Kenncott'. In Anderson, K. and Purcell, S. (eds), *Proceedings, International Conference on Pollution Prevention in Mining and Mineral Processing.* Colorado School of Mines, Golden, Colorado, USA, pp. 100–7.
L. CRAWFORD (1996) 'Clean-up gets underway'. *Financial Times*. 6 June.
R. DIMOCK (1995) 'Kennecott has modern mettle for mining'. *The Salt Lake Tribune*, 16. Apr.
Environment Digest (1996) 'Australian plans for mining code of practice'. *Environment Digest*, 1(5).
C. FREEMAN (1992) 'Values, economic growth and the environment'. In Freeman, C. (ed.), *The Economics of Hope: Essays on Technical Change, Economic Growth and the Environment*. London: Pinter Publishers.
A. P. GRAVES (1991) 'Globalisation of the automobile industry: the challenge for Europe'. In Freeman, C., Sharp, M. and Walker, W. (eds), *Technology and the Future of Europe*. London: Pinter Publishers, pp. 261–82.
P. GRAY (1992) 'The mining industry and its environment'. *Mining and Environment Research Network Bulletin*, Oct., pp. 13–14.

R. ISNOR (1993) 'The relationship between environmental regulation and industrial innovation'. Mining and Environment Research Network, Bath, UK. Progress Report.

M. JACOBS (1991) *The Green Economy: Environment, Sustainable Development, and the Politics of the Future.* London: Pluto Press.

R. JORDAN and A. WARHURST (1992) 'The Bolivian mining crisis'. *Resources Policy*, Mar., pp. 1–20.

D. KOSICH (1995) 'Freeport's Indonesia may prophesy mining's political future'. *Mining World News*, 1(3).

G. LAGOS (1992) *Mining and Environment: the Chilean Case.* Mining and Environment Research Network, Bath, UK. MERN Working Paper Series, No. 23, Mar.

G. LIN (1992) *Environmental Management, Technical Innovation and Sustainable Development of the Non-Ferrous Metal Mines in China.* Paper presented at the Third Workshop of the Mining and Environment Research Network, 14–16, Sept. Wiston House, Steyning, UK.

F. LOAYZA (1993) *Environmental Management of Mining Companies in Bolivia: Implications for Environmental and Industrial Policies Aiming at Sustainable Growth in Low-income Countries.* Mining and Environment Research Network, Bath, UK. MERN Working Paper Series, No. 29, July.

D. H. MEADOWS, D. L. MEADOWS, J. RANDERS and W. W. BEHRENS (1972) *The Limits to Growth: a Report for the Club of Rome's Project on the Predicament of Mankind.* New York: Universe Books.

Metals Economics Group (1994) *Environmental Costs and Issues in Gold Exploration, Development, Production and Reclamation: Metals Economics Group Strategic Report, Volume III.* Nova Scotia, Canada: Metals Economics Group Press.

S. R. MILLIMAN and R. PRINCE (1989) 'Firm incentives to promote technological change in pollution control'. *Journal of Environmental Economics and Management*, 17, pp. 247–65.

A. NUÑEZ-BARRIGA (1993) 'Environmental management in a heterogeneous mining industry: the case of Peru'. Mining and Environment Research Network, Bath, UK. *MERN Research Bulletin*, Dec., pp. 12–13.

OECD (1991) *Environmental Policy: How to Apply Economic Instruments.* Paris: OECD.

L. ORR (1976) 'Incentives for innovation as the basis of effluent charge strategy'. *American Economic Review*, 5(6), pp. 441–7.

R. ROTHWELL (1981) 'Some indirect impacts of government regulation on industrial innovation in the United States'. *Technological Forecasting and Social Change*, 19, pp. 57–80.

R. ROTHWELL (1992) 'Industrial innovation and government environmental regulation: some lessons from the past'. *Technovation*, 12(7), pp. 447–58.

S. SCHMIDHEINY and F. ZORRAQUIN (1996) *Financing Change: the Financial Community, Eco-efficiency and Sustainable Development.* Massachusetts: MIT Press.

J. SKEA (1993) 'Environmental Issues'. In Dodgson, M and Rothwell, R. (eds), *Handbook of Industrial Innovation.* London: Elgar Press.

J. E. TILTON (1994) 'Mining waste and the polluter-pays principle and the United States'. In Eggert, R. (ed.), *Mining and the Environment: Interna-*

tional Perspectives on Public Policy. Washington DC, Resources for the Future.

A. WARHURST (1990) 'Employment and environmental implications of metals biotechnology.' *World Employment Programme International Labour Organisation (ILO)*, Geneva, Switzerland, Research Working Paper, WEP 2-22/SP 207, Mar.

A. WARHURST (1991a) 'Metals biotechnology for developing countries and case studies from the Andean Group', *Chile and Canada. Resources Policy*, Mar., pp. 54–68.

A. WARHURST (1991b) 'Technology transfer and the development of China's offshore oil industry'. *World Development*, 19(8), pp. 1055–73.

A. WARHURST (1992a) 'Environmental management'. In *Mining and the Environment: the Berlin Guidelines*. London, Mining Journal Books.

A. WARHURST (1992b) 'Environmental management in mining and mineral processing in developing countries'. *Natural Resources Forum*, Feb., pp. 39–48.

A. WARHURST (1992c) 'The limitations of environmental regulation: an argument for technology policy to promote environmental management in mining'. *Paper Prepared for the John M Olin Distinguished Lectureship Series in Mineral Economics, Colorado School of Mines*, USA, Nov.

A. WARHURST (1994a) 'Environmental degradation from mining and mineral processing in developing countries: corporate responses and national policies'. *Organisation for Economic Cooperation and Development Publications and Information Centre*, Washington DC.

A. WARHURST (1994b) 'The limitations of environmental regulation in mining'. In Eggert R. G., (Ed.), *Mining and the Environment: International Perspectives on Public Policy*. Washington DC. Resources for the Future.

A. WARHURST and G. BRIDGE (1997) 'Economic liberalisation, innovation and technology transfer: opportunities for cleaner production in the minerals industry'. *Natural Resources Forum*, 21(5), pp. 1–12.

A. WARHURST and L. J. MACDONNELL (1992) 'Environmental regulations'. In *Mining and the Environment: the Berlin Guidelines*. London, Mining Journal Books.

R. WINTERS and L. MARSHALL (1991) 'Where's the recovery in RCRA?: the re-mining of non-coal abandoned mine sites'. *Proceedings of 12th Annual Meeting of National Association of Abandoned Mine Land Programs*, USA, Sept.

J. L. WHITLOCK and T. I. MUDDER (1989) 'The Homestake wastewater treatment process: biological removal of toxic parameters from cyanidation wastewaters and bioassay effluent evaluation'. *Internal Working Paper, Homestake Mining Corporation*, San Francisco.

World Commission on Environment and Development (1987) *Our Common Future*. UK. Oxford University Press.

J. WOMACK, D. I. JONES and D. ROOS (1990) *The Machine that Changed the World: Based on the Massachusetts Institute of Technology 5-Million Dollar 5-Year Study on the Future of the Automobile*. New York, Macmillan.

2 Globalization and Social Responsibility: Issues in Corporate Citizenship

Malcolm McIntosh

'Would you tell me, please, which way I ought to go from here?'
she asked.
'That depends a good deal on where you want to get to' said the
cat.

Alice's Adventures in Wonderland by Lewis Carroll

I INTRODUCTION

Let's start with two very different views of the globalization process.
The first comes from a former television newscaster and managing
director of BBC World Service who now runs a premier London arts
centre. The second comes from a former Clinton administration official and US business consultant.

> Are we content with a world made safe for Mcfood and Waltcul-
> ture and Rupertnews and Tedvision? Will we lie down before the
> high priests of globalization who create a new totalitarianism of
> taste, thought, experience and views using self-serving economic
> arguments decked out in the pseudo-democratic clothing of
> freedom and choice?.... The homogenized global culture snuffs
> out species of artistic endeavor as surely as global enterprise threat-
> ens the bio-diversity of natural species.[1]

The impact of globalization on culture and the impact of culture
on globalization merit discussion. The homogenizing influences of
globalization that are most often condemned by the new national-
ists and by cultural romanticists are actually positive; globalization
promotes integration and the removal not only of cultural barriers
but of many of the negative dimensions of culture. Globalization is
a vital step toward both a more stable world and better lives for the

people in it. . . . For the US, a central objective of an Information Age foreign policy must be to win the battle of the world's information flows, dominating the airwaves as Great Britain once ruled the seas. . . . Americans should not deny the fact that of all the nations in the world, theirs is the most just and the best model for the future.[2]

We live in a global economy. We are all capitalists now, aren't we? The market has triumphed and corporations stride the world looking for new markets, new customers and cheaper resources. But what does this mean? What are the characteristics of the process of globalization that is now taking place? Do we have a shared understanding of the process and the outcome? What are the effects of the global economy on us as individuals, on organizations and business? How can we manage the complexities of the global economy when there is no clear view of the future? What competencies must managers develop for the 21st century in order that our corporations remain profitable and remember their social responsibilities?

There is a feeling that markets are not enough; that capitalism cannot provide for all societies' needs and that business needs to work hand in hand with good government to address environmental issues, distribute wealth more fairly, fight corruption, oppression and human rights' abuses. If business is the principle engine of society it has a clear responsibility not to abuse its new freedom and global role. As Charles Handy says: 'Capitalism is not capable of delivering a good life for all or a decent society. I don't think we should expect it to. It is a means not an end.'[3]

Even international agencies like the World Bank and the IMF, themselves evangelists for the market and trade liberalization, recognize that business cannot be asked to save the planet by itself.

Businesses are becoming aware of a transparent world where producers link with non-government organizations to inform governments, customers, and the media around the world. There is growing pressure on companies to acknowledge their global social responsibilities.

One of the key aspects of the globalization process involves sourcing (procuring and purchasing) from the cheapest market. This can be a major decision for a manufacturing industry, but it can be possible to change sources rapidly for food importers dependent on consistent quality and reliability of supply. The global economy has, of course, existed for hundreds of years and many European empires in the 18th

Box 2.1   Some recent global events for countries and companies

- RTZ–CRA, Britain and the world's largest mining company, and Freeport–McMoRan, an American firm, are attacked by environmentalists and local people for their copper and gold mine in Irian Jaya, Indonesia. They disrupt RTZ–CRA's annual shareholders' meeting and the dissension receives equal press coverage to the company's $2,460 million profit in 1995.[4]
- The International Monetary Fund (IMF) lends Mexico $17,000 million in 1997 to repay foreign creditors. Deregulation of financial markets and privatization had allowed Mexican banks to borrow overseas against local assets. Foreign money poured in, knowing that in the last resort that the IMF would not allow the Mexican economy to go bust. Private speculators were rewarded with high returns guaranteed by an UN agency intent on promoting the free trade. In the end the IMF will force Mexico to repay the debts, but it will not be the overseas speculators who pay, but ordinary Mexicans.[5]
- Prime Minister Mahathir Mohammed of Malaysia calls for exchange rate currency speculation to be outlawed at the IMF–World Bank annual meeting after his country's currency is hit by overseas speculators, as Mexico and Thailand had suffered earlier in the year. The US Treasury Secretary, Robert Rubin, calls for greater globalization of finance, not the reverse.[6]
- In 1997 Nomura, the Japanese finance house, becomes the largest public house owner in the UK paying $1,980 million to GrandMet and Foster's. Along with previous acquisitions this means that that Nomura owns 8% of British pubs, 4,309 'locals'. Founded in 1925, last year Nomura's president, Hideo Sakamaki, was arrested for corruption and fraud in Japan. Along with 16 other directors he admitted to having dealings with *sokaiya* (corporate racketeers). *Sokaiya*, linked to the Japanese Mafia, *yakuza*, had demanded protection money to allow Nomura to operate in Japan. Nomura has since been banned from being involved in certain business areas in Japan by the Japanese government.[7]
- In Paris, France, on 1 October 1997 the French government ordered half the private car owners not to drive due to dangerously high levels of air pollution. This reduces the number of vehicles on the streets from 3 million to 2 million. One week earlier, on 24 September 1997, *The Financial Times* survey of Europe's most respected companies had awarded BP and Royal Dutch/Shell second and eleventh places respectively. BMW came seventh.[8]
- For several months in 1997 Malaysia, Singapore and Indonesia are almost brought to a standstill by smog which forces schools and the airports to close and people to stay at home. The World Health Organization gives a warning of respiratory and heart ailments. The smog is blamed on forest clearance fires which have got out of hand. In 1994 a similar situation arose and was found to be caused by clearing land to produce palm oil and paper and pulp for export.[9]

and 19th centuries were built on trade and cheap overseas sourcing. But the advent of refrigeration early in the 20th century meant an increase in food products, while today vast quantities of perishable fruit and vegetables are airfreighted around the world on a daily basis. The cost to customers in the US and Europe is composed of cheap labour and production costs in the source country plus a substantial proportion representing transport, advertising and packaging costs in the destination country.

One point that needs to be made concerning global sourcing is that money, sourcing and manufacturing may be mobile but that people and environmental resources are normally not. Money is attracted to high returns, manufacturing to low labour rates and sourcing to the cheapest market. While investors, manufacturers and buyers may move on, people remain behind often with the results of careless raw material extraction.

While moving an industrial manufacturing plant from one country to another may be a strategic decision of some magnitude, imposing significant costs, the sourcing of computing and information systems globally has fewer implications for long-term planning. As North America, Europe and some Pacific Rim countries become information and entertainment economies, they find themselves competing with countries such as India, Mexico, Egypt and, in the future, China. It is easier for a small percentage of the population of these countries to tool themselves up intellectually and compete in software systems than it is to set up a car factory.

However the global economy is developing, by flying lilies round the world, money laundering or air-freighting baby corn, figures from the 1997 UN Human Development Report show that the world is becoming polarized between the rich and the poor. That gap doubled between 1981 and 1996. And yet globalization is normally presented, particularly by transnationals and proponents of free trade, as providing benefits to the whole world. The figures presented here show the power and responsibility that global business has in working towards a fair distribution of resources and creating a sustainable world.

Business and government leaders now recognize that governments alone cannot solve problems of poverty and environmental degradation. Let us not fool ourselves that the finger of blame can be pointed at one political system, one industrial sector or one philosophy for the state in which the world now finds itself in at the end of the 20th century.

What is clear is that business operates best in democratic societies

---

Box 2.2   Global business: a different perspective

The trade in illegal drugs
The international trade in illegal drugs, as defined by the United Nations, represents 8% of world trade, is bigger than iron and steel, and second only to the oil business. The UN Drug Control Progamme says that although establishing the size of the illegal drugs trade is difficult, because all activities are outside the law, their estimates are:

- The annual turnover is $4,000 million.
- By contrast annual development aid amounts to $690 million
- In 1996 an estimated 8 million people used heroin, 13 million used cocaine, 30 million used amphetamine-type drugs, 141 million used cannabis and 227 million used sedatives
- World production of coca leaf more than doubled between 1985 and 1996, in the same period opium production more than tripled
- The business is highly globalized with producers and wholesalers able to switch trade routes and markets.

Cracking the launderette
However hard international banks try to monitor the laundering of drug money it is often impossible, although US laws on transparency are increasingly allowing drugs enforcement agencies to track cash flows around the world.

As some of the main recipient countries for laundering drugs money, the US, Britain, Italy, Switzerland and Canada have adopted common codes to help trace funds:

- Banks are required to report suspicious transactions.
- Employees are granted immunity from prosecution for breaking confidentiality laws.
- The sharing of seized assets with other interested governments is allowed.
- Non-banks must meet the same anti-laundering criteria as banks.[10]

---

which operate under the rule of law and with low corruption and with an infrastructure of education, health care and crime prevention provided by central and local government. In other words there has to be an understanding between government, civil society and business that they need each other to build healthy, safe communities.

II   A NEW BUSINESS DEAL

It is on this basis that a new model of global social development is emerging with partnerships between business, governments and com-

---

Box 2.3 **Global business and the state of the world**

- The largest 100 companies have annual revenues that exceed the GDP of 50% of the world's nation-states.
- General Motors has the same annual revenue as Austria.
- Korean motor manufacturer Daewoo, with a workforce of 91,000, has the same annual revenue as Bangladesh, with a population of 116 million.
- Globally 358 billionaires have as much wealth as the poorest 45% of the world's population.
- 359 corporations account for 40% of world trade.
- 5 companies in Britain receive almost 50% of everything the British spend.
- The 12 most important global industries, such as textiles and media, are each more than 40% controlled by five or fewer corporations.
- Ten corporations control almost every aspect of the world-wide food chain.[11]

But:

- Since the mid-20th century the world has consumed more resources than in all previous human history.
- The 48 least developed nations, with 10% of the world's population, have 0.3% of world trade.
- 1,300 million people live on $1 a day or less.
- 160 million children are undernourished.
- 20% of the world's population will die before they are 40 years old.
- 1 in 4 people in the world live in poverty: on less than one US dollar a day
- The world is divided into roughly three classes of affluence, regardless of the wealth of the country they inhabit. The super-rich over-consumers, middle income 'sustainers', and as many people living in poverty. These classes co-exist in most countries. For instance in 1997 the gap between the rich and the poor is the same in Britain as it is Nigeria.[12]

---

munities at its heart. This is what GrandMet means when it says: 'What happens to society matters to us, because it happens to us'[13] and when McDonald's says it has a 'special responsibility to protect our environment for future generations' because 'this responsibility is derived from our unique relationship with millions of customers'.[14] With this understanding these companies can expect that society will monitor their responsible practice very carefully. The trick for companies is to use societal concern for competitive advantage by having an external social audit published to allay customers concerns and provide quality benchmarks for the industry. This pro-active approach has been adopted by GrandMet and BT and is actively being considered by parts of the Royal-Dutch Shell Group and BP.

Box 2.4

We perceive and deal with social issues in a non-traditional manner at Timberland. Timberland doesn't give money to charity. Instead we try to create a return. We integrate the notion of value creation into all our activities. We create values for ourselves as a company, our employees, our shareholders, out customers, the community and the non-profit organizations we co-operate with. The traditional notion of philanthropy is not adequate. It is not smart or wise to approach the social problems of society with the financial leftovers of companies. By integrating our social activities with the sustainability that will see through hard times, and harness business to work in another fashion.[15]

(Timberland, *Community Enterprise*, 1997)

Something radical and exciting is happening when a leading footwear and clothing company like Timberland, recognized around the world, says that there is another way of doing business. The same message is being communicated by McDonald's through its work with the Environmental Defence Fund, when Unilever promotes fish conservation, when Royal Dutch Shell includes references to human rights in its business principles and when Levi Strauss promotes fair trade and ethical sourcing. All these successful companies saw commercial advantage in being seen to care about the world's social and environmental problems.

Global companies like Avon, GrandMet, BT, Shell and Levi Strauss are following in the footsteps of smaller, socially active companies like Ben & Jerry's and the Body Shop and putting corporate citizenship at the heart of strategic planning. They now understand the benefits of a pro-active approach to social responsibility. Being seen to be a responsible corporate citizen is seen as a competitive issue. Companies that have been exposed to global media scrutiny, such as Avon, Shell, and McDonald's, know that the relationship between business and society is changing radically and rapidly. These companies are welcoming the growing rapport with human rights and environmental groups because it sharpens their competitive edge, and makes them stronger players in the global market.

In a global economy companies are now aware that contradictions in their operating practices in one country can expose their markets in another country.

III HOW HAVE WE GLOBALIZED?

- **Telecommunications**. The growth of telecommunications through telephones, faxes and e-mail has made it possible to communicate electronically on a global basis.
- **Monitoring**. The world can be monitored for military movements, rain forest destruction and the effects of climate change. Most of this information is freely available for individuals, governments and business.
- **Business**. Business sources and operates globally.
- **Images**. Images are flashed round the world, from Pepsi's color change in 1996 to the Chernobyl explosion, and the Chinese government's crackdown in Tiannanmen Square in 1989.
- **Disease**. Diseases such as smallpox are eradicated globally while the elimination of the AIDS virus becomes a global talking point and deaths by car accidents are hardly mentioned despite the fact that cars cause more deaths and accidents than many diseases.
- **Ecology**. The sight of planet Earth from space has helped us realize that we share one world and enabled us to understand its fragility and our position in the universe.
- **Connectivity**. We sense that we are connected globally through shopping, travel, resource use, pollution, and the media.

IV THE THREE CHARACTERISTICS OF THE GLOBAL ECONOMY

Movement and Mobility

- Money is free to move, labour is not.
- Capital flows from market to market.
- Exchange rates change constantly.
- Business outsources to the cheapest suppliers or producers anywhere in the world.
- The rush towards an increasingly urbanized world society.
- Women taking up paid work.
- An increase in people working electronically from home at 20 million in 1996, estimated to be some 200 million by 2016.

- The rapid expansion of global business organizations
- The development of global multilateral organizations and non-government organizations (NGOs).

Change

- There is uncertainty about the future.
- A lack of job security is caused by constant restructuring, delayering, and the relocation of companies.
- New products and processes are constantly being developed.
- There is a flood of new information about the world around us, so that just when we think we understand it, it changes.

Wealth Disparity

- The trading blocks of Europe, North America and the Pacific Rim have grown in affluence to the virtual exclusion of other parts of the world.
- Living standards have polarised across national boundaries into three groups of over-consumers, sustainers and the impoverished.

V   INCREASED GLOBALIZATION?

The growth of the global economy is based on the creation of conditions which will increase business activity. These conditions, it is argued, will lead to greater social development alongside increased world trade. There are those who argue that quite the opposite is happening, as the gap between rich and poor is no longer confined to one country versus another, but to the creation of greater disparities of wealth, health and security within every country. In other words the creation of the global economy and the development of flexible labour markets, with business activity drawn to the cheapest labour, could be understood to mean the importation of the worst situations of the developing world—great wealth disparity, massive job insecurity and very low wage rates for a significant proportion of the population.

Those who argue in favour of in increase in global business activity, such as the World Trade Organization, want to see:

- A level playing field for business; which means opposition to industrial subsidies, and an opposition to higher regional regulations in areas such as health and safety, health care and environmental protection.

- De-regulation of industrial sectors, as airlines in the US and buses in the UK have created greater competition.
- Privatization of state services to introduce competition, business efficiency and market mechanisms to areas run on a public service ethic. This has spread as far as the prison systems in the US and UK.
- Structural Adjustment Programmes (SAPs) introduced by the International Monetary Fund for countries with deficits requiring loans from the IMF and the World Bank. These require cuts in government spending, and therefore the provision of public services to balance government budgets.

VI GLOBAL GOVERNANCE

With the partial demise of the state and the growing dominance of business, it has come to the attention of some business leaders that business prospers best under certain conditions. Unless the business concerned is active in trade in illegal drugs, rare animals, arms or counterfeit CDs a World Bank report said that the best conditions for business were to be found in countries where the state was effective. Perhaps the rolling back of the state has gone far enough? Indeed the World Bank report found that countries with virtually ineffective state apparatus and a lack of democracy were the least likely to support a thriving business economy.

In a survey of 3,600 entrepreneurs in 69 countries Latin America, Eastern Europe and Sub-Saharan Africa the vast majority of respondents reported that the authorities failed to protect property and that the judiciary was not honest or consistent. 40% of respondents in these areas of the world said they had to pay bribes to survive, against 15% in more affluent countries.

In other words the institutionalization of the rule of law and fundamental human rights is good for business. The fundamentals of good country governance are a lack of corruption, participation, a lack of civil war, a reliable judicial system and the maintenance of basic infrastructure, including communications, transport, education and health care.

However, there are some transnationals who are charged with operating to lower standards away from their home countries, and from benefiting from lower wage rates, lower health and safety standards and lax compliance with legislation. In recognition of the fact that some countries do not have reliable legal systems or legal aid in 1997 British Law Lords allowed a case to be brought by a Namibian against a transnational company in the UK courts.

The case involves Edward Connolly, a Scottish maintenance engineer, who claims he was poisoned by uranium dust while working for Rossing Uranium, an RTZ subsidiary based in Namibia. RTZ is Britain's and the world's largest mining company. In the House of Lords the presiding judge said: 'This is a case in which, having regard to the nature of the litigation, substantial justice cannot be done in the appropriate forum (Namibia), but can be done where the resources are available.'

One judge, Lord Hoffman, dissented, putting his finger on one of the key global governance issues for transnationals: 'The defendant is a multinational company, present almost everywhere and certainly present and ready to be sued in Namibia.' As *The Financial Times* reported 'if the presence of RTZ in the UK enabled it to be sued here "any multinational with its parent company in England will be liable to be sued here in respect of its activities anywhere in the world"'.[16]

VII   WHAT IS INTERNATIONAL BUSINESS?

Different terms are used to describe companies which operate in more than one country, and it is worth defining what is meant by the different terms. An international company is a company that is based in one country but trades in other countries; a multinational company is a company that may be based in one country but has bases in other countries, for management, manufacturing or distribution; a transnational company is a company that has its headquarters in one country but operates most of the time outside that home country in a number of other countries (Nestlé for instance only has 2% of its operations in Switzerland, its home base). There are also supranationals who appear to recognize no home base and operate in many countries. There are global companies, who may also be transnationals or supranationals, who manufacture different components in different countries to make a final product which is sold globally. It is difficult in these cases to say 'this product is 100% American or Spanish'. Most large automobile manufacturers and electronics companies are now global.

Transnational corporations can be described in various ways, by annual revenue, number of employees, presence in numerous countries and by other criteria. The UN Conference on Trade and Development (UNCTAD) constructs an 'index on transnationality' by looking at a company's foreign assets to total assets, foreign sales to total sales and foreign employment to total employment. On this basis some of the world's largest companies, and some of the best known brands, such as Coca-Cola, McDonald's, Shell and General Motors are excluded. According to UNCTAD's criteria the world's top 15 transnationals are biased towards transnational from smaller countries:

1.	Nestlé	Food	Switzerland
2.	Thomson	Electronics, telecomms	Canada
3.	Holderbank Finacière	Banking	Switzerland
4.	Seagram	Beverages, alcohol	Canada
5.	Solvay		Belgium
6.	ABB	Construction	Sweden/ Switzerland
7.	Electrolux	Electrical equipment	Sweden
8.	Unilever	Food, detergents	Britain/ Netherlands
9.	Philips	Electronics	Netherlands
10.	Roche	Pharmaceuticals	Switzerland
11.	SCA		Sweden
12.	Northern Telecom	Telecommunications	Canada
13.	Glaxo-Wellcome	Pharmaceuticals	Britain
14.	Cable & Wireless	Telecommunications	Britain
15.	Volvo	Automobiles	Sweden[17]

Other indicators of world trade are:

• The US, China and Britain are the largest recipients of foreign investment, the US and Britain being the largest overseas investors.

- Percentage of world trade 1996:

US	13.5
Germany	9.0
France	5.3
Britain	5.1
Japan	7.1
China	2.7
Hong Kong	3.6
Sweden	1.4

Box 2.5   Examples of transnational companies (TNCs)

Levi Strauss, the world's largest clothing company and owners of Levi's, one of the best known global brands, reported sales of $7,100 million in 1996. Being a global company and operating in 40 countries has had major implications for them, leading to the introduction of a code of conduct for their suppliers which guarantees human rights, as well as promoting health and safety standards.

    Neste is one of Europe's largest petro-chemical companies. Based in Finland it operates plants in the US, Canada, France, The Netherlands, Mexico and China. In 1997 they issued their sixth annual environmental report, independently verified by SustainAbility, a leading environmental consultancy. They say: 'Projects aimed at improving customer contacts, quality and work satisfaction support Neste's environmental strategy, which strives towards business operations which are superior to our competitors. . . . Society expects us to be socially responsible in all our operations.'[18]

    NEC, is the world's largest electronics manufacturer, and an original signatory of *Keidanren's* (the Japanese business association) Charter for Good Corporate Behaviour. They say: 'As we approach the 21st Century, the world is rapidly moving from regional interdependence to a global community, due to a highly developed infrastructure and the liberalization of governments. This can truly be called the generation of the 'global village'. . . . NEC will contribute to a sound environment and a livable society through technology that harmonizes with nature and production that is environmentally friendly.'[19]

    Marks & Spencer, the fourth most profitable retailer in the world, describes itself as 'a thriving global organisation operating successfully in many countries round the world.' By the year 2000 it aims to have 120 franchise stores in 32 countries. Its non-UK sales already amount to £1,500 million a year. The company says that despite their higher prices that they live up to their mission of Quality Service Value and 'our customers choose us'. They say: 'We consider ourselves to be part of the communities we serve and actively contribute to creating a more prosperous and self-sufficient society.'[20]

VIII TOWARDS AN ENVIRONMENT RESEARCH AGENDA

Box 2.6 **How globalization is described**

Globalization is being presented with an air of inevitability and over-whelming conviction. Not since the heyday of free trade in the 19th century has economic theory elicited such widespread certainty. The biggest winners have been multi-national corporations. Globalization has its winners and losers.

UN *Human Development Report*, 1997

No society has had so many centers of power as the society in which we now live. . . . Therefore we come back to the old problem of pluralistic society: Who takes care of the Common Good? Who defines it?

Peter, Drucker, '*The New Society of Organizations*',
Harvard Business Review, Sept.–Oct. 1992.

Employers can move abroad, but employees cannot. There is no substantive difference between American workers being driven from their jobs by their fellow *domestic* workers who agree to work 12-hour days, earn less than the minimum wage, or be fired if they join a union—all of which are illegal under US law—and their similarly being disadvantaged by *foreign* workers doing the same. If society is unwilling to accept the former, why should it countenance the latter? Globalization generates an inequality in bargaining power that 60 years of labor legislation in the US has tried to prevent. It is in effect eroding a social understanding that has long been settled.

Dani, Rodrik, *Sense and Nonsense in the Globalization Debate*,
Foreign Policy, Summer, 1997 p. 19.

Box 2.7

As we approach the twenty-first century, a remarkable convergence of political and economic institutions has taken place around the world. . . . Today virtually all advanced countries have adopted, or are trying to adopt, liberal democratic political institutions. . . . In post-industrial societies further improvements cannot be achieved through ambitious social engineering.

Peter Townsend and Donkor Kwabena, *Global Restructuring and Social Policy*, Policy Press, 1996.

We are creating a world that is becoming more deeply divided between the privileged and the dispossessed, between those who have the power to place themselves beyond the prevailing market forces and those who have become sacrificial offerings on the altar of global competition.

David Korten, *When Corporations Rule the World*,
Earthscan, 1995, p. 214

As every country adopts, or is forced to adapt to, liberal economics, it is up to business, as the prime beneficiary, to take the lead in social development that protects the environment and builds just and healthy local and global communities. The evidence is that some businesses have accepted this challenge and are working for their own and the common good. In other words some companies have accepted that there need not be a conflict between profits and shared values. Being socially responsible is a win–win situation. The environment and communities win and business wins. In other words society improves together. What happens to society happens to business, and vice versa.

Acknowledgements

For a development of the ideas in this chapter and a fuller understanding of corporate citizenship please see Malcolm McIntosh, Deborah Leipziger, Keith Jones and Gill Coleman, *Corporate Citizenship: Successful Strategies for Responsible Companies* (*Financial Times*–Pitman, 1998). Malcolm is grateful to his co-authors for their help with this chapter. An on-going series on global corporate citizenship is being developed for publication. Please contact Malcolm McIntosh on email ccumm@wbs.warwick.ac.uk

Notes

1. John Tusa, *The Agony and the Ecstasy Guardian*, 3 Mar. 1996. John Tusa is managing director of London's Barbican Arts Centre and was formerly managing director of BBC World Service and for some 25 years a senior newscaster.
2. David Rothkopf, 'In Praise of Cultural Imperialism?' *Foreign Policy*, Summer, 1997. David Rothkopf is managing director of Kissinger Associates and an adjunct professor of international affairs at Columbia University. He served as a senior official in the US Department of Commerce during the first term of the Clinton administration.
3. Charles Handy *What's It All for?: Re-inventing Capitalism for the Next Century*, *RSA Journal*, Dec. 1996.
4. *The Economist, The Fun of Being a Multinational*, 20 July 1996.
5. Kevin Watkins, *Lender of Last Resort Favors rich*, *Guardian*, 22 Sept. 1997.
6. Larry Elliot, *A Green Light that Means Stop*, *Guardian*, 22 Sept. 1997.
7. Ian King, *Nomura Becomes UK's Biggest Pubs Company*, *Guardian*, 23 Sept. 1997.

8. BBC Radio 4, *The World At One*, 1 Oct. 1997 and *The Financial Times Europe's most respected companies*, 24 Sept. 1997.
9. *The Economist, An Asian Pea-souper*, 27 Sept. 1997.
10. Personal communications, *Guardian*, 26 June 1997; *The Economist*, *'That infernal washing machine'*, 26 July 1997.
11. Paul Ekins, *Wealth Beyond Measure*, Gaia Books, 1992; David C. Korten, *When Corporations Rule the World*, Earthscan, 1995: *New Economics*, Autumn 1996; Will Hutton, *The State We're in*, Vintage, 1996; *1997 Human Development Report*, Oxford University Press.
12. See Alan Thein Durning, *Income Distribution Worsening*. In Lester Brown *et al. Vital Signs 1992–3*, Earthscan, 1992; UN *Human Development Report 1997*, op. cit.
13. Grand Metropolitan, *Report on Corporate Citizenship*, 1997.
14. McDonald's, *Our Commitment to the Environment*, 1996, UK.
15. Ken Freitas, Vice, President, *Community Enterprise*, Timberland, 1997.
16. John Mason, *RTZ Ruling Threatens Other Multinationals, The Financial Times*, 25 July 1997; John Vidal, *See you in Court, Guardian*, 28 July 1997.
17. *The Economist*, 27 Sept. 1997.
18. Neste, *Corporate Environmental Report: Progress in 1996*.
19. NEC, *The NEC Environmental Charter, 1991*.
20. Marks & Spencer, *Environment Report, 1996*.

Further Reading

CANNON, TOM, *Corporate Responsibility*, *Financial Times*–Pitman, 1992.
HAWKEN, PAUL, *The Ecology of Commerce*, Phoenix, 1994.
KELLY, GAVIN, KELLY DOMINIC and GAMBLE, ANDREW (eds), *Stakeholder Capitalism*, Macmillan, 1997.
KORTEN, DAVID C., *When Corporations Rule the World*, Earthscan, 1995.
MCINTOSH, MALCOLM, LEIPZIGER, DEBORAH, JONES, KEITH and COLEMAN, GILL, *Corporate Citizenship: Successful Strategies for Responsible Companies*, Financial Times–Pitman, 1998.
MURPHY, DAVID F. and BENDELL, JEM, *In The Company of Partners*, Policy Press, 1997.
ZADEK, SIMON, PRUZAN, PETER and EVANS, RICHARD, *Building Corporate Accountability*, Earthscan, 1997.

3 Making Domestic Energy Use Visible

Gwendolyn Brandon and Alan Day

Summary

This chapter is concerned with the development of strategies for reducing domestic energy consumption in the UK by providing householders with information and feedback on their energy use. Two experiments are reported; the first being a study of 140 households where six different types of feedback were provided and energy consumption monitored over an entire heating season. One group, that was provided with computer-based feedback, achieved a 15% reduction in energy conservation and it was concluded that to be effective feedback must increase the visibility of energy consumption, be targeted to the socio-economic circumstances of the householder and provide information which is relevant. In the second experiment, which is on-going, an interactive computer display has been developed which monitors energy conservation continuously and presents this information, along with energy advice, to the householders. Twenty of these 'smart' meters have been installed and are being tested over 12 months to establish the levels of energy use reduction that can be achieved.

I INTRODUCTION

Sustainability, will be achieved most effectively if national government policies, aimed at tackling both global and local environmental problems, serve to support local community initiatives. To these ends, Local Agenda 21 has involved local authorities and other organizations in identifying environmental problems and developing appropriate policy responses. Domestic energy consumption represents one area where the links between global problems and individual behaviour are clearly identifiable and so it has become one of the first sustainability issues to be tackled through the combined efforts of national and local government. The year 1995 saw the British government pass the Home Energy Conservation Act, which commits all Local Authorities to achieving a 30% improvement in the energy efficiency of their housing stock in order to reduce national CO_2 outputs.

Given such a policy context and the widespread agreement that reductions in the use of finite fossil fuels are critical if sustainability is to be achieved, two research projects at the University of Bath, funded by the EPSRC Sustainable Cities Programme, have addressed questions related to the best methods for promoting domestic energy efficiency. Previous research, much of it emanating from the US during the oil crisis of the 1970s, suggested that focusing on the development of energy efficient buildings and appliances alone, without reference to the social context and psychological characteristics of the people using them, is not a sufficient way to tackle the problems. Feedback, that is the presentation of information about a person's performance in relation to a given area of behaviour, has been identified by such research as an essential component of policies successful at promoting more energy efficient lifestyles.

The particular success of feedback, as opposed to more general information, has been seen as providing 'both the motivational and informative components necessary for conservation actions to occur' (Stern and Gardner, 1981, p. 330). More recent work (Arvola, 1993) also recognizes the importance of feedback as a starting point in the process of promoting domestic energy efficiency suggesting that feedback serves as a prompt highlighting the disparity that can often exist between people's environmental concerns and their actual behaviour (van Raaij and Verhallen, 1981; Brandon, 1993).

Nonetheless, despite the general consensus on the importance of energy consumption feedback reached by previous research, both theoretical and empirical, it was clear that further research was needed. The issue of what form such feedback should take, or indeed whether the impact of different feedback varies depending on the target audience, were questions that had remained largely unresolved. Furthermore, if these issues had been addressed, it was seldom within a UK context and typically involved field experiments of a very short duration with small samples (Farhar and Fitzpatrick, 1989). For environmental sustainability to be seriously addressed it is imperative that all sectors of society are involved and that any obstacles to environmentally conscious behaviour are identified. It was therefore the aim of this research to use the theoretical context of feedback to provide practical and feasible suggestions about how policy could involve the general public in energy conservation.

II PROJECT ONE: MONITORING THE CITY

The first of the University of Bath's research projects was thus concerned with exploring the relationship between public attitudes towards domestic energy consumption and any related behaviour, specifically in terms of how different feedback might mediate the relationship. Energy use within the homes of most consumers in developed countries has become, to a large extent, an invisible resource, yet in the UK it accounts for 30% of the CO_2 we produce. Quarterly bills or monthly statements provide the only visible record of that consumption by which time the links between specific activities and the energy consumed have been dislocated, a situation described elsewhere as akin to a supermarket not displaying any prices but just giving the shopper a total non-itemized bill at the checkout (Stern and Aronson, 1984). By providing a group of people with energy consumption feedback which is more frequent (importantly reducing the time span between action and consequence) and informative than typical utility bills, the energy people use becomes more visible (Winett and Kagel, 1984; Ester, 1985). The aims of the project were therefore:

- to provide policy relevant findings about what type of information is most successful at encouraging energy efficient actions;
- to see how any such effects relate to existing socio-demographic and structural variables thus enabling the identification of areas where targeted policies could decrease domestic energy consumption.

This approach was designed to allow the researchers to offer explanations about why people have generally failed to respond actively to their environmental concerns and to inform the debate about how policies aimed at making our society more sustainable should be framed, targeted and delivered.

Procedure

In November 1994, 1,000 residents of Georgian properties in Bath were sent letters requesting their participation in the Monitoring the Cities research project. While it is recognized that its architectural heritage makes Bath a particular case, the terraced houses which make up a large proportion of its accommodation have proved a flexible building type and the UK has a significant proportion of its housing stock made up of such buildings. The researchers thought that it was imperative that energy efficiency should be examined in relation to existing buildings given the need, in response to the Home Energy Conservation Act, for energy efficiency measures to relate to all housing stock and all householders. Furthermore, in terms of experimental rigour, by limiting the study to a particular building type the variables relating to building form and construction could be controlled, thus allowing the research to focus more clearly on energy consumption patterns resulting from different occupant behaviours.

A sample of 140 households were finally selected from 178 positive responses with people being omitted on the grounds of not being sole occupants (student houses or flat shares were excluded due to the likely fluctuations in household size and makeup) and for being in residence under two years, as the researchers needed to build a profile of consumption patterns based on historic data. Using criteria that previous research had suggested were relevant to differences in energy consumption, such as number of occupants, their age and tenure status, the sample was sub-divided into seven groups, including one control, with these variables being distributed equally across all groups.

The sample households were visited between January and March of 1995 and all completed an interview-delivered attitude questionnaire designed to establish their existing attitudes towards the environment and towards energy consumption and conservation. A record was made of any conservation activities they already engaged in and surveys were undertaken to gather details about their heating systems, appliances and any energy efficiency measures already in place.

The Dependent Variable

In order to conclude whether feedback had any impact on energy consumption and to be able to evaluate any changes if they occurred, the researchers had to establish pre-feedback consumption patterns and also gauge the capacity any household had for altering their consumption levels. In the first instance this entailed constructing an energy use profile for each household. Consumption figures for gas, normal, peak and off-peak electricity were obtained from the utilities for the two years prior to the survey. These were then weather corrected using data from two local weather stations and the Meteorological Office. In addition there was a six-week energy monitoring period prior to the feedback to provide base-line consumption figures and to enable assessment of the relative energy use of each household.

Secondly, before the potential for change, and thus the relative significance of any consumption changes, could be established, several variables had to be accounted for. Clearly, the presence of certain structural features within a home will have an impact on the energy efficiency of a property regardless of the behaviour of the occupants. For example, good insulation or thermostatically controlled heating systems will help reduce heat loss or waste. Similarly, there are many things that residents can do which are purely behavioural but which can affect their energy consumption and go some way to ameliorating any shortfall in structural factors. All the participating households were thus attributed both a structural and a behavioural 'Potential for Change' rating. The former category was arrived at by assessing the structural energy-saving features present within the homes, such as insulation, lagged tanks, shutters and so on, while behavioural potential was established by awarding points for each energy conserving activity residents already engaged in, such as limiting heating periods, putting on extra clothing rather than a fire and so on.

Feedback

If, as previous research indicates, there is a consensus that feedback does have a role to play in the promotion of energy conservation, then, the University of Bath researchers felt that the research should examine the relative influence of different forms of feedback. Furthermore, as a considerable body of experimental work on feedback already exists, it was decided that the feedback conditions for the Bath research should make reference to the feedback characteristics such work found to be significant. This would permit comparisons across time and conditions to be made and help to refine any conclusions on energy consumption feedback generally.

Three themes were found to dominate previous literature. First, whether feedback contained an element of comparison, either in terms of comparing one person or household to another or by comparing the same person's consumption with their previous performance (Arvola, 1993; Pallak *et al.*, 1980). Secondly, the issue of personal values was repeatedly mentioned; specifically whether economic motives or environmental motives would influence behaviour and therefore whether feedback appealing to one or other of these values would be more successful (Hutton *et al.*, 1986; Axelrod and Lehman, 1993). Thirdly, the issues of how information was presented in terms of how personal or specific it was, was discussed (Stern and Aronson, 1984; van Houwelingen and Van Raaij, 1989)

As a result the original sample was split into three paired groups to reflect these themes. A 'comparative' pair, a 'values' pair and a 'presentation' pair. There was also a control group making a total of seven different groups (see below). Each, including the control, received a Department of Environment (DoE) energy saving information pack prior to the feedback period. The DoE literature constituted the basis of the then government's attempts to address the issue of domestic energy efficiency and the researchers felt that by utilizing it within the fieldwork its relevance to a cross section of the general public could be assessed. Every month between July 1995 and April 1996 different feedback reports were sent to the respective groups apart from the control group who received no further information or feedback for the duration of the project. A standardization process was used to even out periods between meter readings if the number of days covered, as a result of delays or access problems, varied. All the feedback (with the exception of Group 6) was designed so that it could be easily adapted and delivered as part of a nationwide energy policy.

Group 1—Self-versus-others comparison

Group 1 was sent monthly feedback which informed them of their energy consumption for the month and compared it to an average figure based on all the other properties in the project with a similar size and occupancy profile.

Group 2—Self-versus-self comparison

Group 2's monthly feedback was presented as a comparison to their weather corrected consumption for the same period in the previous year.

Group 3—Financial values

This group received monthly feedback which informed them of their energy consumption in both kWh and equivalent monetary value. They were also provided with an information sheet at the start of the project which gave the running costs of a number of household appliances.

Group 4—Environmental values

This group was sent monthly feedback which informed them of their energy consumption in relation to related environmental problems such as acid rain and global warming.

Group 5—Leaflet presentation

Group 5 were provided with a full literature pack at the start of the project that gave them advice on methods of saving energy presented in a generalized written format. Each month, they received feedback that presented their consumption visually by means of simple graphs.

Group 6—Computer presentation

Although not deliverable as easily as other forms of feedback, it was decided to test the impact of a computer-based system as this was seen as a method that could present the most personalized information and was likely to become more widespread in the future. Each property in this group was therefore provided with a personal computer (PC) for their feedback. The PCs contained 3 different programmes. The first required users to input periodic meter readings, at their discretion, which were converted to kilowatt hours of consumption and plotted on a graph which already displayed their own previous year's consumption thus allowing direct comparisons to be made. The second contained a questionnaire on general aspects of energy saving while the third was a directory of energy saving information and advice. This group still received a monthly mailing containing their last month's consumption figures, with a reminder to input these onto their PC, plus a statement about the external temperature for the month (to help them make more accurate comparisons with their historical data).

Group 7—Control

No feedback.

Results and Findings

Tenure and structural energy efficiency

Initial analysis of the relationship between energy consumption prior to feedback and a variety of socio-economic variables found that the type of tenure held by a household related closely to the structural energy efficiency of their homes. The results revealed that Local Authority properties were generally in the middle of the structural energy efficiency scale while mortgaged and properties owned outright accounted for the peaks in structural energy efficiency. Clearly, home owners stand to gain in several ways when they invest in energy efficiency, not only in terms of the improved comfort and savings on fuel costs but also because the value of their property is likely to increase, serving to justify the initial capital outlay.

Local Authority houses were typified by their absence from the lower efficiency properties. This distribution is seen as both an indication of the Local Authority's commitment to maintaining its properties with regards to their energy efficiency and also a reflection of the greater likelihood of tenants in such properties qualifying for subsidized insulation and draught proofing programmes, such as those run by Energy Advice Centres. The properties, in terms of tenure, most likely to be energy inefficient were found to be those that were rented through the private sector. This was interpreted as the probable result of the more transitory nature of private tenants, and thus their lack of incentive for investing in long-term energy efficiency measures, in combination with a situation where landlords might be loath to invest in measures which required capital outlay.

Consumption changes: non-feedback variables

In order to assess what changes in consumption could be attributed to the introduction of feedback it was necessary to establish which factors, if any, related to household energy consumption *before* the research began.

A multiple regression was performed using previous consumption as the dependent variable to ascertain whether income and demographic variables, identified by earlier research as important, were influential in this study. The analysis for the independent variables 'household income', 'age of respondents' and 'number in household' all proved significant and showed, unsurprisingly, that mature households with higher incomes and more occupants consumed more energy. A one-way analysis of variance for all four tenure groups also produced statistically significant differences with those renting from the Local Authority being the lowest consumers. It is likely that this was a function of the relatively high energy efficiency of such properties and also a probable reflection of the low incomes which typified these households. No other variables, including any of the various economic or environmental attitude variables extrapolated from the interview data had any (even marginal) influence on previous energy consumption.

These results contrast markedly with those produced when the dependent variable of the field study was analysed (the total percentage difference between previous consumption and consumption levels at the end of the feedback period). Another series of multiple regressions showed that those variables related to previous consumption levels (income, tenure, age and number of occupants) did not relate to consumption levels *post* feedback. However, the attitude variables, environmental beliefs and predicted personal behaviour, which were unrelated to previous consumption, were found to relate to consumption reductions after the feedback period. Analysis revealed that irrespective of their feedback group, respondents who held pro-environmental attitudes and expressed a pro-active view with regards to their future conservation behaviour, were most likely to have reduced their consumption during the feedback period.

Feedback performance

In terms of differences in energy consumption between the feedback conditions the only group that performed consistently and significantly better than the others was Group 6, the Computer Group. This group was notable not only for the percentage of households which reduced their consumption levels—over 80% compared to the average of 55% among the other groups, but also because the households in this particular group tended to have made greater reductions—on average 15% compared to the previous year's consumption.

The above results refer to total overall fuel consumption, that is both gas and electricity combined in terms of kilowatt hours of energy used. However, it was possible that some households which had reduced their fuel consumption overall might have actually increased their consumption of one particular fuel, but to a lesser extent than the reductions in their other fuel source—resulting in a net decrease in terms of total kilowatt hours consumed. When reductions were analysed on a fuel by fuel basis, the performance of the Computer Group was again markedly different from that of the other feedback conditions. This group was typified by the majority of householders showing reductions on all the fuel sources used in their homes, while such reduction patterns were only the case in about a third of all other households.

It was also noted that, in terms of reduced consumption across all fuel types, those households which recorded a high behavioural potential for change (that is those who were not already engaged in many energy conservation activities) were the most represented.

Discussion and Policy Implications

The results of the experiment clearly demonstrated the impact that feedback could have on domestic energy consumption. In answer to the original research questions it was concluded that the influence of feedback, in a variety of forms, was widespread and cross sectoral. Reductions in consumption were not related to the socio-economic or demographic criteria typically associated with energy consumption levels nor were they related to this sample's pre-feedback energy use. However, while such variables did not pre-determine the likelihood of a household changing their behaviour in response to the feedback, the possession of certain attitude variables did.

Even though the results seem to indicate that, for the majority of people, there is not a direct link between their passive environmental concerns and their lifestyles in terms of energy use, analysis revealed that pro-environmental attitudes were related to the changes in consumption over the feedback period. Whether, as previous work has suggested, the feedback served to highlight this attitude-behaviour gap or whether it simply informed people about links which they didn't know existed, was not concluded.

In terms of policy implications, these findings suggest three areas where efforts should be concentrated if domestic energy conservation is to become more widespread:

(1) Conservation information is likely to be effective if targeted at people with favourable environmental attitudes, providing them with either the psychological prompt or the knowledge to act in response to their concerns. By association, an increase in public environmental awareness, to promote the initial environmental concern, is likely to create an atmosphere where the

public are likely to be more receptive to calls for increased fuel efficiency.

(2) From discussions held with participants at the end of the field-work, it became clear that consumers want customized or par-ticularized advice. General leaflets with often inappropriate information or vague statements about the environment or energy efficiency appear to be inadequate and comments on the effectiveness of the DoE literature.

(3) While feedback generally was effective at promoting changes in behaviour, the only specific feedback form about which the researchers can be confident was that supplied by the personal computers. While it is not a serious policy option to place com-puters in consumers' homes, the success of the computer feed-back is seen to demonstrate the importance of making fuel consumption more visible, both physically and psychologically.

III   PROJECT TWO: SMART METERS

It was the success of the PC-delivered consumption feedback that raised the questions around which the second research project at the University of Bath is centred. With the advent of improved metering technology and the development of 'Smart' meters, the provision of more specific, real-time feedback to domestic users is becoming an increasing possibility. The utilities are already developing meters that enable them to cull considerable amounts of information about indi-vidual households' consumption patterns, all of which helps them to manage the demand for energy thus reducing the need for capital investment. However, the flow of information at present, and the spec-ifications with which such new meters are being developed, is very much in the utilities', as opposed to the consumers', favour. The devel-opment of these meters is best described as supplier led and, as such, may not be particularly useful in the quest to maximize domestic energy conservation through changes at the level of the individual consumer.

The University of Bath, in conjunction with the Centre for Urban Technology at the University of Newcastle upon Tyne, is now devel-oping a 'Smart' metering system to provide households with instant feedback on their consumption. Electronic feedback has been tested elsewhere (van Houwelingen and van Raaij, 1989) but the systems have been very basic, typically only showing the previous 24 hours' consumption and requiring residents to record any cumulative or comparative consumption manually. The smart meter system, however, will enable people to learn about how much energy their various appliances use and to monitor how specific actions in the home affect energy consumption. It has been noted elsewhere (Projects in Partnership, 1984) that people are often unfamiliar with how much energy their homes use and express a desire to know what measures are available to increase energy efficiency.

Another finding from the first project (which confirmed earlier work by Stern and Aronson, 1984) indicated that people want, and respond well to, personalized and relevant information. The metering system being developed will thus be programmable to account for variations in demographic, situational and structural circumstances, taking into account differences in household size, age, heating systems and appliances used, and will respond with specific feedback.

Twenty households have been recruited to the project all of them occupying smaller Georgian or Victorian houses. Family size and life-cycle position varies across the 20 homes and includes retired singles and couples, young families, maturing families, and childless younger couples.

Information Collected

All the householders have been interviewed and completed a questionnaire from which an initial database has been compiled. This contains socio-demographic and economic information, details about heating systems and appliances, and the conservation measures, both structural and behavioural, that each household already engages in. Another database has been compiled containing consumption data for each household for the past two years enabling the meter to be programmed with individual consumption patterns to provide consumers with a comparative element to their feedback.

Meter System and Display Interface

The layout of the elements that make up the smart metering system is shown in Figure 3.1. Data is collected from the gas and electricity meters in real-time and is fed through to the computerized display that also collects information from internal and external temperature sensors. In order to minimize the amount of cabling, the mains power circuit in the house is used to carry the signals to the computer and

Figure 3.1   *Smart metering system schematic*

Gas supply to house   Smart metering system schematic   Electricity supply to house

Gas meter
Type R5

Internal temperature sensor

External temperature sensor   Electricity
transceiver

Siemens
adaptive
electricity
meter

Gas
transceiver

Temperature
transceiver

Socket   Domestic ring main

Socket   Socket

Power supply unit

Display
(Apple Newton)

Serial
transceiver

BT telephone link

RS232 serial data link

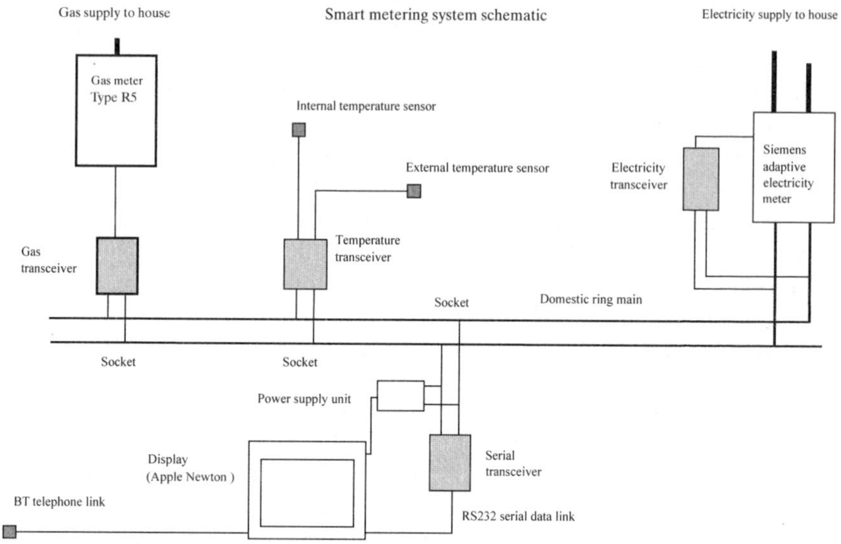

thus gateways are required in order to convert the signal to and from the mains system.

A central element of the new metering system is an attractive and highly visible display. This is the interface for the householder and provides feedback on consumption along with information on how that consumption might be reduced. The display and processor is an Apple Newton MessagePad 2000 which is wall-mounted in a prominent position in the house, normally in the kitchen. This has a touch sensitive screen some 125 × 85 mm in size and allows consumption to be shown either as kilowatt hours or as the equivalent cash value. There is also be an option that resembles a vehicle odometer where consumers can reset the counter to zero and then monitor consumption over a set period. For example, they might choose to reset it at the same time each morning and thus monitor gas and electricity consumption on a daily basis.

As the computer retains information on previous consumption, it is also possible for householders to make comparisons with the amount of energy they have used in previous weeks and even set themselves targets which they can monitor graphically. The Newton is programmed with energy advice in the form of a 'tip of the day' along with more detailed information which can be accessed using the touch screen. This information is customized to suit the profile of the household to ensure that consumers only receive relevant tips. As each Newton is connected to the university using a modem it is possible to update tips in line with current events, such as unseasonably cold weather, and also monitor householders' consumption and their interactions with the Newton.

In order to simulate the effect of variations in energy tariffs, each householder is given a number of points at the start of the experiment. These points have a cash value which can be redeemed at the end of the study and householders will be presented with messages, delivered via the Newton, which will set them targets which, if complied with, will result in bonus points. For example, they might receive a message saying that if they use 20% less energy in the coming week than they did in the previous week they will receive 200 bonus points. It is likely that, as pricing policies become more sophisticated, energy providers will offer supplies on the basis of flexible tariffs which they will use to manage demand. This part of the experiment will allow the researchers to investigate the likely impact of such a development.

The meters will stay in place for one full year after which all participants will be revisited and interviewed about their opinions of and responses to the smart meter display.

IV CONCLUSIONS

The research on energy consumption feedback at the University of Bath has clearly shown that, in addition to those factors earlier work found to be important in the promotion of public energy conservation, there are three key elements which feedback needs in order to successfully precipitate changes in peoples' energy consumption behaviour: visibility, relevance and targeting.

The results of the first project (Monitoring the City) demonstrated the importance of both the physical and psychological visibility of energy consumption. Feedback utilizing a physical presence in the home, in this case in the form of an interactive personal computer, served to increase the visibility of a household's energy use and then provided the prompt for people to adopt energy conservation measures.

By making the feedback more relevant to individual household circumstances, it was found that people's awareness of their personal energy related behaviour could be increased. People, naturally enough, were disinclined to extract information that was relevant to them from information that was general and indiscriminate. The possession of favourable attitudes to the environment, while not related to pre-feedback consumption levels, was found to predispose people to taking energy conservation measures once the feedback was introduced. It seemed to highlight inconsistencies between their attitudes and behaviour and, when coupled with relevant conservation information, provided the stimulus for them to act.

With regards to differences between feedback forms, one specific type of feedback was consistently associated with widespread and significant reductions in energy use: computer-delivered feedback. It would seem that the PCs encapsulated the key elements of how to successfully raise the visibility of energy consumption through physical prominence and personal interaction.

These findings are being explored further in the next stage of the project where personalized and relevant information is being delivered automatically to people's homes using a computer-based feedback system.

In this second stage of the project these 'Smart Meters' are being used to deliver those things which the first stage of the project identified as being important: high visibility, relevance and targeting. The wall-mounted Newton provides the visibility, the data being collected is real-time and therefore always relevant, and the energy advice is directly targeted to the circumstances of the household. Twenty households are having Smart Meters installed and their energy consumption is being monitored over an entire year in order to establish what levels of energy reduction can be achieved using this system.

Although the current installations are relatively expensive at around £1,500 per household there is no doubt that mass production could bring the unit costs down to around £300 which would begin to make the system viable, particularly if it were offered as an incentive by energy suppliers seeking to increase their market share. Alternatively, the interface could be developed to include other household services such as home shopping and banking thus making it more attractive as a consumer product.

Although significant energy savings can be achieved by upgrading the physical fabric of buildings, there are alternative ways to promote conservation. People's behaviour also has a significant impact and can completely undermine the physical improvements if it is inappropriate. It is important, therefore, to develop mechanisms through which attitudes can be changed and appropriate behaviour, which reinforces the physical improvements, adopted. From this research it seems that the most appropriate way of doing this is through measures that provide direct feedback and allow people to take control of their energy use. This provides them with the means of translating sympathetic environmental attitudes into positive action.

References

A. ARVOLA (1993) 'Billing feedback as a means to encourage household electricity conservation: a field experiment in Helsinki.' In Ling, R. and Wilhite, H. (eds), *The Energy Efficiency Challenge for Europe*, Proceedings of ECEEE, Denmark, June 1993.

L. AXELROD and D. LEHMAN (1993) 'Responding to environmental concerns: what factors guide individual action?', *Journal of Environmental Psychology*, vol. 13, pp. 149–59.

G. R. BRANDON (1993) *'Factors affecting the relationship between expressions of environmental concern and related actions,'* unpublished PhD thesis, Imperial College, London.

P. ESTER (1985) *Consumer Behaviour and Energy Conservation*, Dordrecht: Reidel.

B. FARHAR and C. FITZPATRICK (1989) *Effects of Feedback on Residential Electricity Consumption: a Literature Review*, USA: Solar Energy Research Institute.

R. HUTTON, G. MAUSER, P. FILIATRAULT and O. AHTOLA (1986) 'Effects of cost related feedback on consumer behaviour: a field experiment approach', *Journal of Consumer Research*, vol. 13, pp. 327–36.

M. PALLAK, D. COOK and J. SULLIVAN (1980) 'Commitment and energy conservation.' In Bickman, L. (ed.), *Applied Social Psychology Annual*, vol. 1, Sage.

Projects in Partnership (1994) 'Breaking through the barriers to energy saving in the home', *Interim Report, Energy Saving Trust*.

P. C. STERN and E. ARONSON (1984) *Energy Use: the Human Dimension*, New York: Freeman.

P. C. STERN and G. T. GARDNER (1981) 'Psychological research & energy policy', *American Psychologist*, vol. 36, pp. 329–42.

J. H. VAN HOUWELINGEN and W. VAN RAAIJ (1989) 'The effect of goal-setting and daily electronic feedback on in-home energy use', *Journal of Consumer Research*, vol. 16, pp. 98–105.

W. VAN RAAIJ and T. M. VERHALLEN (1981) 'Household behaviour and the use of natural gas for home heating', *Journal of Consumer Research*, vol. 8, pp. 253–7.

R. WINETT and J. KAGEL (1984) 'Effects of information presentation format on resource use in field studies', *Journal of Consumer Research*, vol. 11, pp. 655–67.

4 Green and Ethical Investment: Can It Make a Difference?

Alan Lewis and Craig Mackenzie

Summary

There are a growing number of unit trusts available in the UK with respectable 'green' credentials. People can invest money with some confidence that companies with poor pollution records will be avoided. The authors raise the general question, relevant in the study of economic psychology, of how positive green attitudes can be more effectively married to appropriate action, whether 'green/ethical' investors are special people, whether they are prepared to put their money where their morals are, and ultimately whether or not it will make any difference. The authors base their analysis on a survey of green/ethical investors ($N = 1,146$), focus group studies and telephone interviews. Policy options and research agenda are considered including tax incentives to do the 'right' thing and more active engagement on the part of green unit trusts in improving corporate environmental performance.

I INTRODUCTION

It is now commonplace for opinion surveys to record large majorities of the general public showing concern about environmental issues, depletion of the ozone layer, South American rain forests and so on (Angus Reid Group, 1992). The battle to raise environmental consciousness has largely been won, but is this enough? A branch of psychology, social psychology, studies, among other things, attitudes, values and opinions and how these dispositions are related to behaviour. To cut a long scholarship story short, the evidence shows that the 'right' attitudes however widely held, do not necessarily produce the 'right' behaviour: just because most people care about the environment does not mean that we can be confident that the same people will take direct action to protect it.

There are several dimensions to this apparent inconsistency, for instance, what opportunities exist for people to translate their beliefs into action? Bottle banks are now familiar sights in many cities, and people certainly use them, but what else can ordinary folk do? Tunnelling below roads and airport runways is for cranks and eccentrics however 'loveable' 'Swampy' and the rest may be. 'Ordinary' people have jobs to go to, family responsibilities, a 'life'.

'Green' products of one sort or another are prominent on supermarket shelves, announcing recycled paper packaging, ozone-friendly sprays and fully biodegradable washing powder but they have not always had the impact that was hoped for.

Contemporary school children are very knowledgeable about rain forests, ecology, global warming but will they be prepared to make lifestyle changes, to have aspirations at variance with other generations? When these young people look to the future it seems highly probable that they will want the same things: they will want to own a car, have plenty of money to make desired purchases. The same can be said for the rest of us; the rain forests are one thing but we have not really brought these problems 'home'; we are still in love with the motor car and hardly anyone considers household energy consumption, to take just one example, as harmful or wasteful.

It would be wrong to underplay cultural and ideological dimensions. Evidence is mounting that those to the political left are more concerned about the environment and are more likely to take action than those of the political right; the latter having more confidence in the ingenuity of scientists and technologists to solve the problem (Cotgrove, 1982; Kilbourne *et al.*, 1999). Of course these ideological distinctions are much less clear cut now both in North America and the UK; other authors preferring to talk of the dominant social paradigm (DSP). DSP can be translated as measuring support for existing institutions, the law, government and capitalism. It seems that in the USA it is usual among the young to have allegiance both to the DSP and environmental concern; a belief that environmental problems can be tackled without political and economic upheavals, that the appropriate structures are already in place. It appears, that in some European countries at least, this relationship is reversed: there is a belief that if environmental problems are to be tackled seriously one cannot carry on as before because there are too many vested interests among politicians and business lobbies (including the car and utility lobbies) to bring about fundamental change. This is not an opportunity for smugness among Europeans. While some North Americans may want to have their environmental cake and eat it, many Europeans might recognize that current institutions are ineffective but still want to pass the buck to 'them' (the politicians, leaders in commerce) to do something about it.

II ECONOMIC PSYCHOLOGY AND GREEN INVESTING

Economics and psychology are very different disciplines, yet it is the view of the authors that some kind of combination, an economic psychology, is highly appropriate for the study of environmental problems (Lewis *et al.*, 1995). In psychology it is generally accepted that people have a variety of motives: people do things because it may increase their self-esteem, or because the action in question may be favourably viewed by others, out of obligation, or because they believe it will help them 'grow' as a person, even because some inner voice tells them it is morally the correct thing to do. In economics on the other hand the preference is to explain in the 'objective' world of supply and demand, of scarcity and the efficient distribution of resources; little attention is paid to the 'inner world' of the economic mind, accept in the assumption that people are self-interested utility maximizers. Psychology can be criticized for ignoring economic influences on behaviour and economics for ignoring work that could illuminate the black box that is the decision-maker (this is a complex debate and interested readers should consult Marr and Raj, 1983).

So an economic psychology approach would stress not only environmental attitudes, values and beliefs but also the prices people face in putting their beliefs into action. Policy-makers in a sense already do this, (although may be unfamiliar with the terms): people believe that using unleaded petrol is a good thing, but would they have converted to it had there not been tax incentives to do so? Encouraging people to behave responsibly through financial incentives is a 'win–win' situation; it is our belief that neither moral commitment (or sympathy) nor financial incentives on their own are likely to work, but together they form a powerful combination (there are certainly examples, and car tolling comes to mind, which might back-fire [sic] producing resentment rather than approval).

Intellectually economic psychology is itself driven by a desire to understand how 'ordinary' people make sense of and manage their mixed motives, which might pull them in opposite directions.

We have painted rather a dull and pessimistic picture of environmental consciousness as 'mere' rhetoric and of little consequence, all is not doom and gloom: there are people who choose to invest their savings paying attention to environmental issues as well as the familiar financial considerations of return and risk.

Green investing is an apt case study in economic psychology which has both theoretical and policy implications. Are people prepared to take a loss in order to put their money where their morals are? Are green investors special people? Will it do any good? Could it/should it be encouraged by tax incentives?

III   A BRIEF HISTORY OF GREEN INVESTING WORLDWIDE AND IN THE UK

Green or environmental investment has emerged as an element of ethical or socially responsible investment. The idea that non-financial criteria should be used when considering one's investment is a venerable idea. The question of the ethics of investment goes back at least to Biblical times and the prohibition in Jewish law on usury. John Wesley, founder of the Methodism, emphasized in the 1700s that the use of money was the second most important subject of New Testament teaching. He gave four consecutive addresses entitled: 'Earn all you can'; 'But not at the expense of conscience'; 'Not at the expense of our neighbours' wealth'; 'Not at the expense of our neighbours' health' (Jacob, 1996). These themes have influenced ethical investment since then. However, what is currently recognized as ethical investment—the practice of investing funds on the basis of a set of ethical criteria—is more recent. In the first half of this century a number of Church organizations had adopted various ethical prohibitions on certain kinds of investment, particularly those which have become known as 'sin stocks': alcohol, tobacco, gambling and arms. In the US in 1928, the Pioneer fund was set up, with a policy of deliberately screening certain investments on ethical grounds (Harrington, 1992). However, the current interest in ethical investment began in the late 1960s. A number of church investment funds started to explore ways of avoiding investments in companies operating in South Africa or involved in the Vietnam war. In 1970 the first ethical mutual fund, the Pax World Fund, was established (Harrington, 1992). This was shortly followed by the Dreyfus Third Century Fund in 1972. In the UK, the Church Commissioners of the Church of England had imposed some ethical restrictions on their investments since 1948 but like the US, it was not until the late 1960s that discussion began about ethical unit trusts.

The first application to set up an ethical mutual fund in the UK, in 1973, was rejected by the Department of Trade, because it considered there to be a conflict between the demands of capital and the demands of conscience. However, a successful attempt was made in the early 1980s, and the first UK ethical fund, the Stewardship Unit Trust was established by Friends Provident in 1984.

Ethical unit trusts work by adopting an ethical investment policy with a series of detailed criteria specifying which areas of company activity they will avoid on ethical grounds, and which areas they will favour. Negative criteria have evolved to include avoidance of companies involved in weapons, tobacco and alcohol production, gambling, human rights abuses, health and safety breaches, intensive farming, nuclear power, ozone depletion, greenhouse gas emissions, pesticides, water pollution, tropical hardwoods, and third world concerns. These criteria are applied to the stockmarket leaving a list of companies that are acceptable for investment. Most funds also use positive criteria specifying areas of company activity which are endorsed by the fund. Positive criteria are sometimes used to select companies to target for preferential investment, and sometimes to balance against some of the smaller ethical failings of companies and so to pass them as acceptable. In addition to this screening process, some ethical funds also have a policy of talking to companies to persuade them to adopt more progressive policies.

The initial UK ethical trusts did not have substantial coverage of environmental issues. However, in 1988 Merlin Fund Management launched the Merlin Ecology fund. This was the first screened investment fund to focus explicitly on environmental issues. The initial proposal for the fund said:

Many investors are causing, albeit unwittingly, serious damage to our planet by investing in companies that are harming the environment. At the same time the growing realisation of the problems of pollution and resource depletion has led to the emergence of a number of companies who are actively seeking remedies and making technological advances in pollution control and environmental management. We believe that demand for the products and services of such companies is rapidly expanding, and will contibute to do so for the foreseeable future. In addition there are a growing number of companies which have adapted their products and or production processes to minimise their adverse impact on the environment.

'If your care about the environment, this unit trust is specifically for you', Merlin Ecology Fund, 1988

Merlin later became the Jupiter Ecology Fund. In 1989 the Jupiter International Green investment Trust was launched. Subsequently 6 green funds have been launched in the UK.

- Eagle Star Environmental Opportunities Trust (June 1989)
- TSB Environmental Investor Fund (June 1989)
- Clerical Medical Evergreen (February 1990)
- Cooperative Insurance Society Environ (May 1990)
- Barchester Best of Green (June 1991)
- Commercial Union Environmental Trust (April 1992)
- Sun Life Global Portfolio Ecology Fund (December 1992)

However, while these funds have explicitly marketed themselves as 'green' or 'environmental', other funds, like Friends Provident Stewardship, which market themselves as 'ethical' rather than 'green' now have rather similar investment policies. There are two reasons for this. Green funds, by and large, have adopted a range of ethical policies, on, for example, arms, tobacco and gambling. On the other hand, many ethical funds have adopted a range of environmental criteria, spurred on by the launch of environmental funds, and by the greatly increased public interest in environmental issues in the late 1980s. This means that at the level of the investment policy it is not possible to make firm distinctions between ethical and environmental funds. One exception is the Eagle Star Environmental Opportunities Trust, which claims not to be an ethical trust at all, but aims to share in the success of environmentally progressive companies. This is prudent given that Eagle Star is owned by one of the big four tobacco companies, BAT. However, for the most part we can ignore the presentational differences between ethical and green funds, and consider the whole ethical investment market at 'environmental' for our present purposes.

This market is large and growing fast. In the UK, as of 1 September 1997, there was £1.6 billion of assets managed by 34 ethical/environmental funds, on behalf of perhaps 100,000 investors. Assets under environmental management have doubled in the three years to September 1997. Green funds are growing twice as fast as the unit trust sector as a whole. While this size and growth is impressive, it is still small as a proportion of the overall stockmarket. Green funds account for perhaps 1% of the unit trust sector, and 0.15% of the UK stockmarket as a whole.

IV   A SURVEY OF GREEN/ETHICAL INVESTORS

Are green/ethical investors 'peculiar' people? From a survey of 1,146 ethical investors conducted by the authors the demographic profile is not so different from 'ordinary' investors: they are predominantly middle-aged (and older), middle-class, with above average earnings. The differences are apparent when looking at sex, occupation and membership of societies and religious groups: 46% of our respondents were female; 31% were employed in the education sector; 14% in health (and as a comparison only 4% and 5% respectively were from the manufacturing and retail sectors); 16% were members of the Labour party, 7.7% Liberal Democrats: there were more members of the Green party (2.4%) than the Conservative party. Religious allegiances were well represented by the Church of England (16.1%) and the Society of Friends (10.3%). Among the charities the most popular for membership were The National Trust (36.6%), Amnesty International (31.1%), Friends of the Earth (27.6%) and Greenpeace (26.4%).

These investors have already taken important steps to reflect their convictions where their green and ethical investing form a part of a 'lifestyle' package. This package though is not so neatly tied up with sealing wax and string: to our surprise the majority of green/ethical investors have at best 'ordinary' even 'non-ethical/green' investments at the same time investing on average, about 31% of their portfolios in green/ethical unit trusts. Investors are clearly committed to the green/ethical investments they do have, almost 90% stating they would retain the same proportion invested in this way even when the level of green/ethical investment produced a 2% annual underperformance compared to 'ordinary' investments. Even more striking is the finding that 59% of our respondents would retain their level of green/ethical investment with a 5% annual underperformance which constitutes a significant financial sacrifice.

It is clear that these people are prepared to put their money where their morals are, but not all, or even the majority of their money. Their mixed portfolio is an example of moral and financial pragmatism. This willingness to sacrifice return is not a function of the proportion invested in environmentally acceptable ways, but rather a function of the strength of investors motives, for example in their commitment to avoiding companies that are doing harm, encouraging those who are making a positive contribution to society as well as more general concerns about nuclear power, animal testing, the international sale of armaments, the degradation of the ozone layer and the problems of the Third World.

The majority of green/ethical unit trusts are based on the avoidance of 'bad' companies with poor pollution records and so on, but far from all socially responsible investors feel that this is enough and across several questions in our questionnaire between 70% and 86% agree that they would like to see green funds take a more active stance by, for example, campaigning publicly for companies to adopt better policies (70%); or by contributing actively to debate about corporate ethics and the development of public policy (80%).

V WHAT DO GREEN INVESTORS THINK THEY ARE DOING?

In order to address this question 49 green investors took part in focus groups and a further 20 were interviewed over the telephone:

(1) They are avoiding investment in companies doing the most damage to the environment. There are a number of companies who are regularly targeted by environmental pressure groups, including Shell, BP, RTZ, and Hanson. It is not a coincidence that these companies are all in messy extractive industries. The fact that these high profile 'sinners' are avoided by green and ethical funds is one important reason why green investors choose green funds.

(2) They are supporting more progressive companies: green investors believe that at least some of their money is invested in leading environmental companies. They may therefore believe that they are in some sense importing corporate efforts at environmental improvements.

(3) Green investors hope that green funds are actively lobbying

companies to make less pollution and invest in energy efficiency and cleaner technologies.

(4) Green investors may also feel that they are 'making a statement', aligning their investments with their values, doing their bit to make a better world.

As well as these highly laudable motives participants were not shy in admitting that a good deal of the explanation for their investment decisions were based on their need to 'salve their consciences' as well. They needed to make money, especially in the longer term in the form of capital growth, but they were uncomfortable about how this might be achieved, although by definition they were not prepared to avoid the Stock Market altogether. It was not uncommon for participants to mention their religious beliefs and their other social and environmental beliefs and how these directed their behaviour.

Why then, we inquired, do people hold green and not-so-green investments at the same time? Their replies made us rethink our rather simplistic classification. Participants told us that while they feel commitments to future generations as a whole they also were driven to bequeath to their own immediate family, that this motive, on occasions, outweighed other considerations and was, in a sense, just as ethical as their other motives.

This was not the exclusive explanation and some were mildly, but only mildly disconcerted with the question, pointing out that their 'unethical' investments were legacies and were not entirely 'owned' by them or because green investments were relatively new to the market they had not, as yet, got around to changing them. We got the feeling that 'spare cash' or 'unanticipated windfalls' were much more likely to be socially invested than funds 'earmarked' to cover retirement and future health provision.

Many of our interviewees also gave money to charity but were quite clear that green investing was not of this type and even when returns were low were more effective ways of encouraging environmentally acceptable enterprises as they were concerned that charitable aid alone might lead to inefficiency and dependency.

Watery metaphors and similes often came to our participants' minds. Their investment decisions were mere 'drops in the ocean', but like 'water dripping on stones' they would wear away the hard core of unacceptable corporate practice. There was a feeling that they were part of a movement that was growing and as more and more people invested in this way businesses would have to take stock. There was

also a belief that green investments would, by some subjective definition, bound to be sound in the long run, a belief also held in the financial pages of broadsheet newspapers (Winnett and Lewis, 2000).

VI THEORETICALLY/REALISTICALLY CAN IT MAKE A DIFFERENCE AND IF SO HOW?

There are three ways in which it is suggested that green funds can
make a difference: culture, share boycotting, engagement.

The first and weakest way in which green funds may be able to
make a difference is that their mere existence may contribute to a
process of cultural change. The fact that there is £1.6bn of assets in
the City screened on environmental grounds, gives visibility to green
issues in the financial markets and in corporate boardrooms. The disproportionate amount of coverage these funds achieve in the serious
newspapers enhances this effect. However, this contribution to cultural change is not measurable.

A second argument why green investment might make a difference
is what you might call the 'share boycott' argument. The argument
goes like this: if green investors boycott particular shares, they will,
because of the law of supply and demand, drive down the price of
those shares; and because company directors care deeply about the
share price, they will accede to the demands of green funds. But do
green funds have the power to move share prices? There are a number
of reasons for thinking that it is very unlikely. One reason is that green
funds in the UK constitute well under 1% of total value of the stockmarket. The fact that such a relatively small pool of assets refrains
from buying shares in, say, Shell, will not make a noticeable difference
to its share price. Secondly, even if green funds were much larger, their
behaviour in the market place will tend to be compensated for by
other non-ethical actors. While boycotting companies may work in
consumer markets, there are strong reasons for believing that it
cannot work in capital markets.

The third argument why green investment might make a difference
builds on the 'active engagement' model. In the US, active engagement, or shareholder activism, has become a routine part of investment activity within certain sectors of the investment market. Since
the early 1980s a wide coalition of 275 Protestant, Catholic, and Jewish
institutional investors have been engaged in shareholder activism

through the Interfaith Center on Corporate Responsibility (ICCR), directed by Tim Smith. The first church-sponsored shareholder resolution was filed by the Episcopal Church in 1971 with General Motors on their involvement in South Africa (Smith, 1992). Perhaps the largest co-ordinated shareholder activist campaign in recent years has been the attempt to persuade companies to sign up to the Coalition for Environmentally Responsible Economies (CERES) Principles.

The main CERES objective was to persuade companies to sign up to a set of Principles (initially known as the 'Valdez' Principles, after the Exxon Valdez disaster) about corporate environmental practice. To this end, in 1989, 20 shareholder resolutions were filed by church investors working with ICCR, and also by CalPERS and the New York City Retirement System, to ask companies to disclose their compliance with the Principles. The resolutions received an average vote of 12.5% (Bavaria, 1992). Between 1991 and 1994 an average of 40 resolutions a year were filed on the Principles (Hoffman, 1996). Each year some resolutions were withdrawn because the company concerned agreed to negotiate with CERES. By 1995, over 50 companies had endorsed the CERES principles, most of them companies with well-known environmental concerns or commitments—such as Ben and Jerry's and the Body Shop. However, over the last few years some very large companies have also been persuaded to sign, including General Motors, Sun Oil, ITT and Polaroid. Furthermore there is evidence to suggest that the CERES programme has been influential on the environmental practice of a number of other substantial companies, even though they did not adopt the principles (Hoffman, 1996).

While CERES may not have achieved the obvious goal of persuading a large proportion of corporate America to sign up to its Principles, this was never its only purpose. As Joan Bavaria, co-chair of CERES, says 'What we're after is culture change and forging relationships. Our goal is not to become an institution but to be part of the process' (Hoffman, 1996). It is likely that, albeit indirectly, CERES has had a significant effect on the culture of environmental practice of a wide range of large corporations in the US and Europe. While it was not solely responsible for changes in environmental policy, it served as an important catalyst. Hoffman claims that 'CERES's influence in the business environment cannot be disputed . . . the changing landscape of corporate environmentalism bears CERES's mark.' In consequence we can conclude that the large scale, long-term active engagement which CERES embodies can be a effective means of promoting environmental change.

Active engagement in the UK is not widely pursued, at least by green funds. There are a few examples of shareholder activism on social or environmental issues by investors—on the extraction of peat by Fisons plc; on consumer issues with Yorkshire Water; on various issues concerning RTZ; on the Third World debt and the banking sector; on director's pay and British Gas; and most recently on external social and environmental auditing and Shell. Most of these activities have been organized by the institutional shareholder lobby group PIRC (Pensions Investment Research Consultants).

Jupiter Ecology, the first UK green fund, has lead the way in developing the active engagement approach in the UK (though the 'green team' which founded this fund moved to NPI Global Care in 1995 making that fund a leader in this area). This activity, however, is small scale compared to the US. Actually, many of the rest of the green and ethical funds do not pursue any kind of engagement with the companies in which they invest. This means that currently green funds do not have a great deal of impact on raising the standards of corporate environmental practice. However, the US success story indicates that if they did adopt active engagement in a more concerted way, green funds might reasonably be expected to make a significant difference.

VII TOWARDS AN ENVIRONMENT RESEARCH AGENDA

Public opinion is already right behind many aspects of the environmental movement. What is required are more opportunities for people to take action. One of the most productive ways of bringing about a greater cohesion between beliefs and action is by combining some of the more useful contributions from psychology and economics: an economic psychology. Taking green and ethical investing as an example it is clear that at least some people are prepared to take a financial loss in order to turn their environmental beliefs into environmentally positive action. But these are people who have already taken action of various other kinds as well, by joining political parties and environment pressure groups. In order to draw in greater numbers and make green investing more effective there may be a case for favourable tax treatment for certain kinds of ethical and green investing; in particular one might see an extension of the favourable treatment of venture capital investment trusts to investment in companies engaged in environmental and social development (although government 'intervention' in these markets is almost always controversial).

Investing in these funds must do more than merely salving the consciences of individuals and there are some in the field (including one of the authors, Craig Mackenzie) who believe that the way forward is through active engagement, best suited to encouraging corporate change. Investors might be particularly attracted to funds that they deem likely to 'be making a difference', it must be remembered however that there can be an unease among financial service companies, financial advisers, and investors themselves when financial matters and 'politics' embrace one another too closely: there will always be a market for those driven by considerations of risk and return alone.

In psychology it is generally believed that in many contexts positive reinforcement is more effective than punishment. When considering what economists would refer to as the 'externality' of pollution, the most direct route would be to prosecute and fine companies with poor pollution records; where the law/government might be in direct conflict with industry and one which might simply encourage companies to become more devious and hooded. It is argued here that the less direct but more positive means of encouraging greater corporate morality can be achieved by green investment (albeit aided by tax incentives) where companies are reinforced for doing the right thing, and where improved practice is rewarded by increased investment from identified green investment funds.

As a continuation of this line of research we would recommend the following:

(1) A study investigating the effectiveness of the engagement approach from green/ethical unit trusts to corporate ethics, perhaps with an international perspective.
(2) Psychologists, economists, financial experts and management scientists working together to assess the effectiveness of tax incentives for green investing. Might a green stakeholder (or should it be shareholder?) make for a better world?

Bibliography

ANGUS REID GROUP (1992) *Canadians and the Environment,* Vancouver, BC: Angus Reid Group.

J. BAVARIA (1992) 'CERES and the Valdez Principles'. In Kinder, P., Lydenberg, S. and Domini, A. (eds), *The Social Investment Almanac*, New York: Henry Holt.

S. COTGROVE (1982) *Catastrophe or Cornucopia*, New York: Wiley.

A. DOMINI and P. KINDER (1986) *Ethical Investing*, Reading, Mass.: Addison Wesley.

J. HARRINGTON (1992) *Investing with Your Conscience*, New York: Wiley.

A. HOFFMAN (1996) 'A strategic response to investor activism', *Sloan Management Review*, Winter 1996, pp. 51–64.

C. JACOB (1996) 'Address to the UKSIF AGM', unpublished manuscript.

B. KILBOURNE, S. BECKMANN, A. LEWIS and Y. VAN DAM (1999) 'Differences in environmental attitudes: a multi-national examination of the dominant social paradigm', unpublished manuscript.

A. LEWIS and C. MACKENZIE (1999) 'Support for investor activism among U.K. ethical investors' (forthcoming, *Journal of Business Ethics*).

A. LEWIS and C. MACKENZIE (2000) 'Morals, money, ethical investing and economic psychology' (forthcoming, *Human Relations*).

A. LEWIS, C. MACKENZIE, P. WEBLEY and A. WINNETT (1998) 'Morals and markets: some theoretical and policy implications of ethical investment'. In Taylor-Gooby, P. (ed.), *Choice and Public Policy*, London: Macmillan.

A. LEWIS, P. WEBLEY and A. FURNHAM (1995) *The New Economic Mind*, New York and London: Harvester Wheatsheaf.

C. MACKENZIE (1993) *The Shareholder Action Handbook*, Newcastle: New Consumer.

C. MACKENZIE (1997) 'Ethical investment and the challenge of corporate reform', PhD Thesis, University of Bath.

W. MARR and B. RAJ (1983) *How Economists Explain*, Lanham: University Press of America.

T. SMITH (1992) 'Shareholder activism'. In Kinder, P., Lydenberg, S. and Domini, A. (eds), *The Social Investment Almanac*, New York: Henry Holt.

A. WINNETT and A. LEWIS (2000) ' "You'd Have to be Green to Invest in this": popular economic models, financial journalism and ethical investment' (forthcoming, *Journal of Economic Psychology*).

5 Environmental Education and Teacher Education: a Critical Review of Effective Theory and Practice

William Scott and Christopher Oulton

Summary

This chapter begins with an exploration of the links between environmentalism and education, and looks at the argument that an educational response is important as a means of combating environmental problems. The chapter then traces the development of the notion that the introduction of environmental education into teacher education programmes should become the priority of priorities, and examines a variety of responses to this injunction. In particular, the work of UNESCO–UNEP is critiqued and the outcomes of two influential European projects examined in some detail. Finally, more recent developments in both Europe and North America are charted. The chapter ends with a call for environmental education to become more focused around the work of schools and universities, and for a research agenda to be developed to support this.

I THE ENVIRONMENT AND EDUCATION: THE CASE FOR ENVIRONMENTAL EDUCATION

Although many countries have long traditions of educational approaches which have required learners to spend time studying outwith schools, both in local communities and farther afield in what might be termed 'the environment', it is only since the late 1960s that the term environmental education has found common usage. Originally, such traditions of outdoor education, nature studies and urban studies, which were supported by voluntary, local, national and non-governmental organizations, had a range of aims relating to lifestyle, health, nature conservation and the transmission of particular sets of values. It was from these beginnings that environmental education emerged when the nature and seriousness of what is known as the environmental or ecological crisis became clearer.

Over the past 40 years or so, theories of environmentalism and environmental education have developed, with theorists in the latter field linking their work to emerging ideas in the former. Prominent among those modelling the emergence of environmentalism was O'Riordan (1981, 1983, 1989, 1990) who made distinctions between anthropocentric and ecocentric perspectives which he saw as having reductionist and holistic tendencies, respectively, and which might be labelled with varying shades of 'green-ness' (O'Riordan, 1983). Sterling (1990, 1993) offers an extended critical commentary on O'Riordan's work; see also Sessions (1974), Sandbach (1980), Cotgrove (1982), Milbrath (1984), Pepper (1984), Naess (1989), McCormick (1989) and Fox (1990) for a discussion of further issues relating to environmentalism.

Sterling (1993, p. 83) in developing O'Riordan's (1983) model of environmentalism, distinguishes two holistic/ecocentric stances (dark-green and red-green) and two reductionist/technocentric (light-green and non-green) stances, and this dichotomy represents the prevailing way of viewing the models, although O'Riordan was the first to point out that none of the perspectives were particularly firm or impermeable; and even the simple division by Sterling has now been critized (Gough *et al.*, 2000).

There have been two principal and quite distinct perspectives to the argument that an educational response needs to be made to the environmental crisis. The first perspective, which might be termed 'reformist', has a purpose of ensuring that the threats to all lifeforms and ecosystems on the planet are understood more widely than they are now and, perhaps more crucially, takes an ecological perspective in doing this in order to show how organisms interrelate with each other, and with their environments, throughout the planet's ecosystems with the aim of altering human behaviour and thus lessen the damage to the planet. The second perspective, which might be termed 'reconstructionist', has the quite different purpose of influencing the worldviews that individuals and social groups espouse with the aim of changing the nature of the current socio-economic system in order to lessen damage to the planet.

The reformist perspective involves showing how human activities—personal, social and particularly industrial—have been implicated in contributing to the crisis, and how changes in human activities might be wrought in order to effect some amelioration of the problems. Thus, this argument can be couched in terms of the need to provide appropriate information about how 'planet Earth' works, and about human and environment interactions within it, and also in terms of a critical examination of the current crisis and what might be done to address the issues. Clearly, in terms of a school, there is great scope for contributions to this perspective from across the curriculum and within different age groupings. The main thrusts here will be cognitive, and much mainstream formal environmental education has been focused around these ends. Though it might be thought that such an approach might command widespread support because of its potential to raise learners' awareness of the problems and their urgencies, this is far from the case. The emphases of this perspective have been widely identified by socially-critical theorists such as Huckle (1993) and Fien (1993a,b) as being entirely associated with the reductionist and technocentric parts of O'Riordan's model, i.e. the so-called 'light-green' or 'non-green' stances that assume, respectively, that change and benefit can be achieved by gradualist approaches, working within the grain of society, or that the need to change the problem is overstated. Being thus designated, they are then viewed as part of the problem rather than being able to make a contribution to any solution. Huckle (1993, p. 56), for example, argues for 'real, or ecologically sustainable development' which he contrasts with 'education for environmental management' (p. 63) which he sees as 'consistent with the more general process whereby education and schooling are being "conformed" in the interest of hegemonising the prevailing "commonsense" ideology of the dominant class'.

The reconstructive perspective sees the reformist position as essentially representing a conservative response to a serious problem which requires a radical set of solutions. Its purpose is to influence the way that individuals, groups and societies view the world and their role within it, and to influence their individual and collective values with the purpose of fostering more holistic understandings of the interconnectedness of humanity and the planet. There have been two perspectives here. One, the so-called 'dark-green' ecocentric and anti-humanist view, essentially puts *homo sapiens* no more than on a par with other species in a relationship which Capra (1982, p. 62) sees as 'the oneness of all living forms and their cyclical rhythms of birth and death, thus reflecting an attitude towards life which is profoundly ecological'. This perspective sees the need for fundamental change in the ways that the majority of human societies organize themselves in order to achieve what might be termed ecological justice. The other is a so-called 'red-green' view that takes a socially-critical, humanist, anthropocentric perspective (Gough *et al.*, 2000) which sees the need for radical and/or revolutionary change in the way societies govern themselves in order to achieve social justice. Any educational thrust here, will need to be in the affective domain. Over the years, more rhetoric than mainstream school time has been focused around these ends, although there is clearly scope for contributions to these issues across the curriculum and ages; finding appropriate curriculum niches is, however, seen as problematic because of the overtly controversial (i.e. political) nature of the argument.

Much of the initial development work on environmental education was carried out through a series of international initiatives which were focused around the reformist perspective. These included: the United Nations Conference on the Human Environment (UNCHE) in Stockholm, 1972, the International Environmental Education Programme (IEEP) and the Belgrade Charter—a global framework for environmental education, 1975; and the First International Conference on Environmental Education at Tbilisi in 1977 hosted by Unesco (Unesco, 1972, 1976, 1977a,b,d, 1980; 1992; Unesco–UNEP, 1992; World Commission on Environment and Development, 1987; IUCN/UNEP/WWF 1980, 1991). This work continues today both internationally (United Nations, 1997) and on a local scale, worldwide (for example, Wiltshire Agenda 21, 1997). A recurring theme within such developments has been the need for teacher education to be prioritized.

II   ENVIRONMENTAL TEACHER EDUCATION: OUR INTELLECTUAL CHALLENGE

One of the research foci of the team of environmental educators within the University of Bath has been the development of an enhanced understanding of the theory and practice of pre-service teacher education programmes in relation to the wider role of environmental education within schools. The intellectual challenge for us comes directly from the international development of ideas relating to the theory and practice of environmental education as developed over the past years and which we have set out, above. Specifically, it comes from the Unesco–UNEP assertion that teacher education has to be the 'priority of priorities' and the national and international initiatives which have followed from this; for example, developments which have outlined objectives and guiding principles for environmental education in Europe which identified initial teacher training and in-service training as a priority for its schools (Gayford, 1991).

The notion that the education of teachers should involve an environmental dimension was first agreed at the 1971 IUCN conference in Switzerland where representatives of over a hundred countries recognized that teacher education forms one of the most important and significant aspects in the development of environmental education programmes, and recommended that:

- the training of teachers provide them with essential basic knowledge of ecological facts and an adequate background of sociology and its relationship to human ecology;
- efforts should be made to develop in teachers a critical awareness of environmental problems to enable them to provoke responsible attitudes concerning environmental matters in their pupils;
- environmental conservation is recognized as an essential part of teacher training and that developments started in pre-service training should be continued by in-service training;
- as teacher training in environmental education involves the use of many techniques and methods, all prospective teachers should be given training in the use and evaluation of pedagogic methods, including those relating to inter-disciplinary approaches and team teaching.
- media banks should be established at national and international level for the exchange of information, training and teaching materials.

(IUCN/UNEP/WWF 1972,3)

These proposals embody an essentially transmissive model of training and teaching: one in which information is given to teachers who, in their turn, then pass it on to learners through particular pedagogies, seemingly irrespective of the school context, local conditions or learner-predispositions, or particular educational or developmental needs.

In 1975, the Belgrade Charter refined the details of the model. It saw teachers as one of the 'principal audiences of environmental education', and called for well-designed programmes to be developed with the aim of educating teachers. The Charter also gave rise to a series of initiatives over the next few years which attempted to develop principles for the design and development of a teacher education environmental education programme of study. A series of Unesco meetings across the world in 1976/77 supported such developments and resulted in a publication *Needs and Priorities in Environmental Education: an International Survey* (Unesco, 1977c) which documented needs as seen across countries and at different educational levels.

It was, however, the Unesco inter-governmental Tbilisi conference in 1977 which was the seminal event in bringing the crucial role of teacher education to the fore. Here ministers unanimously agreed that environmental education should be an obligatory part of both pre- and in-service programmes and considered a 'priority activity' (Unesco, 1978, p. 5). One of the principles set out at Tbilisi was that there was a need 'to strengthen ordinary pre-service and in-service training programmes [for teachers] aimed at making them capable of including an environmental component in their teaching activities'. Tbilisi called for national programmes of action, and Resolutions 10 and 11 of the Tbilisi declaration said that such programmes should include a basic level of training (in-service and pre-service) which would enable teachers to incorporate environmental education effectively into their activities.

Such proposals beg a large number of questions, for example:

- What exactly is 'essential basic knowledge of ecological facts' and 'an adequate background of sociology and its relationship to human ecology'? How basic is 'basic'? How adequate is 'adequate'?
- What constitutes a 'critical awareness of environmental problems'?
- How do teachers 'provoke responsible attitudes concerning environmental matters in (their) pupils'? Indeed, should they? Which attitudes are these? Whose attitudes are they? Who decides?
- Which 'pedagogic methods' are 'most appropriate'?
- How do you 'strengthen ordinary pre-service and in-service training programmes [for teachers] aimed at making them capable of including an environmental component in their teaching activities'?
- How 'basic' is the 'level of training' required (by Unesco) which

would 'enable teachers to incorporate environmental education effectively into their activities'? And are we sure that teachers are enthusiastic about doing this?

The need for coherent programmes of environmental education in all aspects and phases of education has also been argued persuasively by a number of national governments, supra-national bodies and non-governmental organizations. However, the literature provides a recurring testimony to the failure to achieve desired goals, and nowhere yet does environmental education appear to have been introduced in a consistent or coherent fashion into pre- or in-service teacher education programmes. See, for example, Selim (1972), Unesco (1978), Mishra *et al.* (1985), Williams (1985, 1992), Unesco–UNEP (1992) and QBTR (1993) for a 25-year commentary on the difficulties. Although there undoubtedly are a number of factors at work here, Scott (1996) suggests that four issues in particular are significant: (1) the lack of a shared conception of what environmental education within pre-service courses is, or ought to be, and the lack of an agreed set of goals; (2) the absence of an understood and agreed pedagogical approach to working with novice (trainee) teachers in this aspect of the curriculum; (3) too great a focus by teacher educators on knowledge transmission without sufficient, if any, use of research-led, learner-focused, or interdisciplinary approaches; (4) an insufficient consideration of the professional competencies needed by novice teachers for their work in schools. In short, there is a large degree of uncertainty about what constitutes effective teaching and learning in the field. There is also the chronic problem that there is only a small number of schools where environmental education programmes are sufficiently well-developed to act as exemplars of effective practice which can be drawn upon in teacher education programmes. As a consequence, even where novice teachers have a disposition to incorporate environmental issues into their work with students, the school in which they are placed for teaching experience may not provide a suitable context within which those they can experiment with approaches to environmental education.

Scott (1994) has also examined how diversity within pre-service courses across the European Union militates against environmental education's being featured more often, or incorporated more readily, with courses for novice teachers. Scott claims that there are a number of separate elements to such diversity within programmes of pre-service teacher education, i.e. programme organization, practice in terms of working with novice teachers and with schools, the interpretation of environmental education found within courses, the readiness and ability to incorporate environmental education within courses, and the opportunity to deliver environmental education goals through pre-service courses. For example, where pre-service teacher education remains rudimentary, the opportunities for any environmental education intervention are severely limited. Scott suggests that these are inevitable outcomes of a range of factors which are to do with both environmental education and initial teacher education, and that they are multi-layered.

Scott (1996) has identified a number of such layers of diversity: the national context; universities; pre-service courses themselves; course managers and teams; lecturers; partner schools with which courses work; individual novice teachers. For example, where there is a compulsory national curriculum with detailed programmes of study, it is possibly to have a top-down specification of environmental education, although this in itself will not guarantee effectiveness at school level. In terms of both teachers' and novice teachers' varied experiences in schools and of out-of-school activities, it is probably that they themselves will have diverse backgrounds of environmental education and quite differing awarenesses and understandings of environmental issues and environmental education, and also a marked variation in their empathy towards the latter's development. Thus, there is likely to be diversity in teachers' and novice teachers' perception of the nature of curriculum and any role for environmental education within it, and also of their understandings of the ways in which they might make a personal contribution to environmental education through their own curriculum specialism.

It follows from this that any pre-service provision has to deal, not only with a multiplicity of issues with both cognitive and affective dimensions but also is likely to have to begin from a varied base of experience and commitment. The vision of the virtuous developmental cycle whereby environmentally educated graduates prepare to become teachers who, in their turn, contribute to environmental education in schools is far from being realized. Any implementation theories which are developed need to be grounded in the complexities of such diversity.

Over the years there have been numerous attempts to respond to these questions, not only by practitioners and researchers such as ourselves but also by environmental education organizations nationally and internationally. Prominent among these has been Unesco–UNEP which undertook development work in the late 1980s and published a series of papers around the turn of the decade: Unesco–UNEP's 'International Environmental Education Programme'. As noted earlier, their ideas were summarized as the priority of priorities (Unesco–UNEP, 1990)

Unesco–UNEP begins by defining the desired result of environmental education training programmes for teachers as 'foundation competences in professional education' and 'competences in environmental education content' (see Appendix 1). In many ways the ideas in this paper are informative, but they are not particularly useful, and there are severe limitations in their practicable helpfulness. We have written extensively about these issues elsewhere (Oulton and Scott, 1995). In summary, our main objections are that the Unesco–UNEP proposals are:

- in some important regards, inappropriately conceptualized;
- not specific enough for progress to be made;
- too heavily focused on ecology, to the exclusion of other disciplines;
- orientated to ends (product) at the expense of means (process);
- lacking in reference to the realities of how innovation occurs;
- insufficiently differentiated between requirements for in-service & pre-service provision, and between the needs of primary & secondary schools;
- not focusing on environmental sustainability issues (Unesco–UNEP, 1992).

We are not alone in finding such approaches problematic. Tilbury (1992), for example, has charted in considerable detail such calls over a 20-year period and has discussed a number of models which emerged in the 1980s. Like us, she finds them all wanting in some regard: Stapp *et al.* (1980): much too limited in their appreciation of the need for a strategic approach to change within teacher education institutions; (Hungerford *et al.* (1988): insufficiently contexualized in the realities of the curriculum; Marcinkowski *et al.* (1990): too specialized; Unesco–UNEP (1990): overly content-focused.

Our critique of the Unesco–UNEP ideas stems from our own research in Bath in conjunction with colleagues across the European Union. While all of the critical points made above are in need of attention, in our own research we have chosen to focus on what we feel are the two important questions as far as pre-service programmes are concerned:

- What limits do you realistically need to place on the focus and ambitions of such programmes?
- What should the priorities be for pre-service programmes in terms of organization, content and approach, given the limited state of environmental education within such programmes currently?

The European Union-funded EEITE programme (Brinkman and Scott, 1994, 1996) has explored these issues, and we have commented on the outcomes in detail elsewhere (Oulton and Scott, 1995). One outcome of EEITE's work was a series of organizational principles which it was felt should underpin the work of pre-service courses. These, which were set out in the form of course aims, programme elements, and didactics characteristics which might inform the work of pre-service teacher education, are shown in Appendix 2.

III   DEVELOPING THESE IDEAS

The ideas derived from the EEITE project were further developed by Scott (1996) who established a tentative implementation theory which can inform the work of teacher educators working with novice teachers in their initial professional development. Scott examined the work of EEITE in conjunction with that of the OECD-funded ENSI project (**EN**vironment and **S**chools **I**nitiative) which was a co-operative curriculum development programme based in primary and secondary schools. He took the key ideas from both EEITE and ENSI (see Appendix 3), arguing that the resulting implementation theory was a criterial frame of process skills and values which can inform and guide the inclusion of environmental education within pre-service teacher education programmes. The implementation theory is shown at Appendix 4.

Scott (1996) states that the theory takes the following as axiomatic: all teachers, irrespective of any phase or academic specialism have a role to play in working together to further the goals of environmental education and, therefore:

- All novice teachers' pre-service professional development should contain elements which allow the development of appropriate awareness, understanding, commitment and pedagogical and organizational skills.
- These elements need to be integrated within initial teacher education courses in interdisciplinary and other ways which are syn-

ergistic with the aims of the pre-service courses themselves.

- Within an agreed set of overall aims, pre-service courses have a double agenda of a set of goals determining tutors' work with novice teachers, and a further set of related, yet distinctive, goals determining novice teachers' own work with students.
- Each of these agenda is realized through an active collaboration between tutors and novices with students, teachers and community representatives.
- The development of novice teachers' pedagogical and organizational skills relating to environmental education develops by means of a process of systematic reflection about its purposes and its application in practice; becoming a practising, environmentally educated and environmentally educating teacher is thus a process of researched action.

It is clear that the theory places a premium on certain kinds of experience and activity. The theory valorizes a particular way of working with novice teachers and certain interactions with tutors, and espouses a double agenda of a set of goals determining tutors' work with novice teachers, and a further set of related, yet distinctive goals, determining novice teachers' own work with students. The theory calls for certain modes of course organization to be privileged; for example, a process of systematic reflection about its purposes and its application in practice.

It is through these processes that values are made explicit by being lived and brought into the open through reflection. This implementation theory is part of 'a criterial frame of process aims and values' (Elliott, 1995, p. 7) which embodies an understanding of what experience is important and why. This begs the question: what values are these? In terms of producing the theory, value stances were instrumental at each iteration of the process. These included acknowledging that novice teachers:

- are facilitators of learning;
- have development needs themselves, and seeing the desirability of novices learning alongside teachers and students by collaborative working within the school and its community;
- need to focus on pedagogy as a priority, and on their own development of professional awareness, understanding and competence;
- need to engage with professional issues through the development of their work with students in the curriculum; the focus here needs to be on co-operative professional and curriculum development;

- need to be able to use the locality as a source of learning for themselves and for students with, in novices' case, that learning being focused on the development of pedagogical skills, and the need for collaboration in order to pool expertise and effort, and to work together;
- have the dual need to learn themselves, and to develop their own pedagogical skills while supporting and facilitating student learning, and that it is important that managers, teachers, novice teachers, students and others work co-operatively together to create opportunities to work with and for the community;
- need to be able to develop pedagogical skills which allows 'environmental experience' to become the stimulus for, and focus of, work with students which embodies action-taking.

But the theory ignores novice teachers' needs to investigate, reflect on and evaluate a range of what one might term established, 'external' theories of teaching and learning in order to help them to make best judgements about the most appropriate theories to develop and use within their own particular contexts.

IV   OTHER RECENT DEVELOPMENTS

In the meanwhile, in the United States, developmental work has continued in isolation from European and Australian experience, principally through the offices of the North American Association for Environmental Education (NAAEE) in its National Standards for Environmental Education Project which is linked to the national education goals scheme in the USA. In this process NAAEE produced 'Developing a Framework for Environmental Educator Performance Standards' which is 'an examination of existing guidelines or competencies for environmental educators as well as a synthesis of these guidelines that will help direct the development of environmental education performance standards. The purpose is to provide an overview of the relevant literature and spur deliberation' (NAAEE, 1994b, p. 62). The introduction to the project states:

Educators' Performance Standards

The standards developed for this portion of the project will focus on the skills and knowledge needed by environmental educators. It is anticipated that many of the standards will reflect 'good practice' regardless of the setting or content. They will attempt to describe what environmental educators need to do to provide meaningful environmental education experiences. The standards will be used to inform the design of pre-service and professional development programs for environmental education.

(NAAEE, 1994a, p. 5)

It must be stated explicitly what the standards will not do. The standards are voluntary, and therefore do not define a national curriculum. Although standards outline the core ingredients for quality environmental education, they do not prescribe how EE will be taught at the state or local level. Educators, community members and parents will continue to develop locally appropriate curricula, using the standards as guidelines against which they can monitor the quality of their children's education.

(NAAEE, 1994a, p. 7)

The guidelines which have been drawn on here go back to the IUCN's 1971 ideas, but are little more than a repetition of work done on teacher education by US environmental educators. Although they draw on projects other than those of Unesco–UNEP, and offer a synthesis of the guidelines (p. 88 on), they compound rather than overcome the basic objections to the Unesco–UNEP ideas which we voiced earlier. As the authors point out: 'Looking at the various guidelines, no particular structure or model for the development of environmental education performance standards is really evident. Some of the guidelines are simply listings of desirable attributes' (NAAEE, 1994b, p. 88). Authors then go on to describe how the ideas have been structured by the NAAEE working group: 'Although perhaps a bit arbitrary, the environmental educator competencies have been divided into six themes: (A) Learner Knowledge and Skills Base; (B) Educational and Psychological Foundations; (C) Environmental Education Foundations; (D) Instructional Methodologies; (E) Learning Environment; and (F) Assessment' (ibid., p. 88). Each of these is then briefly examined, and the detailed information about concepts and skills that environmental educators have to command is expressed in over 60 statements taking up almost six A4 pages.

Taking only four of these statements shows the enormity of the task envisaged:

- understand the basic components of societal systems;
- a knowledge of how natural systems work, as well as of how social systems interface with natural systems;
- methods for conserving resources (ecomanagement);
- promote and model curiosity, excitement, wonder, and imaginations as habits of mind.

Some of our earlier critique of the demands made by the Unesco–UNEP ideas remain here, and any concerns which have been alleviated are more than outweighed by an additional worry about the enormity of the task facing anyone who wishes to cover this ground. It reads like a life-time's work just to become qualified as a teacher. Surely all this misses a fundamental point which is that, just as all teachers will be teachers of literacy in their native tongue, irrespective of any specialism, so must they be teachers of environmental literacy through their own (subject) specialism(s). These recent NAAEE proposals once more promote the notion of the 'renaissance wo/man', the polymath who could staff a school single-handedly. It is hard to see where such teachers will come from, and they are certainly not in schools at the moment. The only realistic course of action seems to be to encourage and help teachers do what they *can* do within their own area of expertize, working with them on the ground to extend this.

The production of more and more lengthy lists of what teachers *should* know, understand and do before they can even begin to make a contribution seems increasingly misplaced. Even though the notion of a national standard was clearly little more than a set of guidance which was voluntary in all but name (NAAEE, 1994a, p. 7), there were immense difficulties. Roth (1997, p. 28) notes: 'the goal of establishing national environmental education standards is apparently falling of its own weight' and 'the NAAEE has already backed off and changed the name of their publication to *Environmental Education Materials: Guidelines For Excellence*' (Simmons, 1997). All of this continuing stress on detailed learning outcomes in environmental education is ironic given the degree of doubt and concern about the appropriateness of current conceptions of environmental education and the outcomes which might derive from it, including from within the USA. For further discussion, see Jickling (1995, 1997), Simmons (1996) and Wade (1996), Gough (1997), Oulton and Scott (2000).

The standards/excellence debate and its critique, though not conducted primarily around environmental education/teacher education issues, fits the context well; when it comes down to it, whether it is a question of environmental education in schools with students, or environmental education within universities with would-be environmental educators, the question is whether to focus on the learning/teaching process itself as we ourselves, and commentators such as Wals and van der Leij (1997) would argue or on the outcomes of that process. If you believe that educators need to be free to make decisions about outcomes in collaboration with learners, parents and communities, or with novice teachers, schools and teachers in relation to teacher education, then you will always see an emphasis on process as being more important than one on outcomes. We argue the logic of this position elsewhere (Scott and Oulton, 1999) where we argue for a multiplicity of approaches, carefully and communally deliberated on, to deliver the educational goals deemed appropriate and necessary by communities. Such a strategy will be cross-disciplinary in that it will be informed by a combination of traditions and ideological persuasions which *together* will offer more than any one of them could alone. These same arguments apply to environmental teacher education.

V TOWARDS AN ENVIRONMENT RESEARCH AGENDA

Much of this chapter has been concerned with setting out a research agenda which can explore emerging realities of environmental education within pre-service teacher education programmes in universities and schools. In doing so, a view of the relationship between practitioners and researchers, and practice and research has emerged which runs counter to much received wisdom within environmental education. Over the past 20 years or so, much of the research energy and effort which has gone into environmental education has been focused on refining definitions and/or bolstering rival ideological positions, in order to persuade practitioners that one view was superior to another. This process has represented, on the one hand, an emphasis on 'product' (in respect of pre-determined outcomes in terms of knowledge, skills, and behaviours) in order to effect changes to how we live our lives within the present socio-economic system, and, on the other, an emphasis on 'process' (in respect of, pre-determined values and attitudes) in order to stimulate a change in that system. Both approaches have tended to see schools as recipients of research outcomes, and/or as places where researchers set the agenda for *in situ* investigations.

However, as we hope that this chapter makes clear, there are other views which have implications for research agenda. Where responsibility for deciding what constitutes the most appropriate environmental education in particular contexts (whether school or university) rests, in part at least, with those working in that context, our argument is that responsibility for determining the research agenda also rests, in large part, with those working in the context. Thus, there is an imperative for those in universities to devote part of their research effort to working in partnership with practitioners in schools in order, firstly, to support such research, and secondly to help link that endeavour with their own research work and with related developments elsewhere. Thus, our continuing work with schools and teachers constitutes one aspect of an evolving agenda which focuses on the range of issues which we have already discussed, and our complementary research activity continues to address these questions from different perspectives. In particular, we are interested in questions of how best can one work with teachers and schools in order to aid their own deliberations about working definitions of environmental education, to enhance their implementation strategies, and their choice of appropriate pedagogies. We are also interested in questions of how best to link pre- and in-service provision and to integrate these with wider initiatives such as Agenda 21.

Further Reading

In addition to the sources cited in the chapter, the following may be of interest:

D. C. CANTRELL (1993) 'Alternative paradigms in environmental education research: the interpretive perspective'. In Mrazek, R. (ed.), *Alternative Paradigms in Environmental Education*, Troy, OH: NAAEE.

N. GOUGH (1993) 'Narrative inquiry and critical pragmatism: liberating research in environmental education'. In Mrazek, R. (ed.), *Alternative Paradigms in Environmental Education*, Troy, OH: NAAEE.

H. R. HUNGERFORD, T. L. VOLK, B. G. DIXON, T. J. MARCINKOWSKI and P. C. ARCHIBALD (1988) 'An Environmental Education Approach to the Training of Elementary Teachers: a teacher education programme', *International Environmental Education Programme; Environmental Education Series No. 27*, Paris: Unesco–UNEP.

T. J. MARCINKOWSKI, T. L. VOLK and H. R. HUNGERFORD (1990) 'An environmental education approach to the training of middle level teachers: a prototype programme', *International Environmental Education Programme; Environmental Education Series No. 30*, Paris: Unesco–UNEP.

M. MARIEN (ed.) (1996) *Environmental Issues and Sustainable Futures—a*

Critical Guide to Recent Books, Reports and Periodicals, Bethesda, MY: World Future Society.

I. ROBOTTOM, 'Matching the purposes of environmental education with consistent approaches to research and professional development', *Australian Journal of Environmental Education,* 8 (1992), pp. 133–46.

J. C. SMYTH, 'Environment & education: a view of a changing scene', *Environmental Education Research,* 1(1) (1995), pp. 3–20.

S. STERLING (1996) 'Education in change'. In Huckle, J. and Sterling, S. (eds), *Education for Sustainability,* London: Earthscan.

R. J. WILKE, R. B. PEYTON and H. R. HUNGERFORD (1987) 'Strategies for the training of teachers in environmental education', *International Environmental Education Programme; Environmental Education Series No. 25,* Paris: Unesco–UNEP.

References

F. G. BRINKMAN and W. A. H. SCOTT (eds) (1994) *Environmental Education into Initial Teacher Education in Europe (EEITE) 'the State of the Art',* ATEE Cahiers No. 8, Brussels: Brussels: Association of Teacher Education in Europe.

F. G. BRINKMAN and W. A. H. SCOTT (eds) (1990) 'Reviewing a European Union initiative on environmental education within programmes of pre-service teacher education', *Environmental Education Research,* 2(1), pp. 5–16.

F. CAPRA (1982) *The Turning Point,* London: Wildwood House.

S. COTGROVE (1982) *Catastrophe or Cornucopia: the Environment, Politics and the Future,* Chichester: Wiley.

J. ELLIOTT (1995) *Action Research and School Initiatives in Environmental Education,* paper presented at the 1995 EERA Conference, Bath: University of Bath, Sept. 1995.

J FIEN (1993a) *Education for the Environment: Critical Curriculum Theorizing and Environmental Education,* Geelong: Deakin University Press.

J. FIEN (ed.) (1993b) *Environmental Education: a Pathway to Sustainability,* Geelong: Deakin University Press.

W. FOX (1990) *Towards a Transpersonal Ecology,* Boston: Shambala.

C. GAYFORD, 'Environmental Education: a Question of Emphasis in the School Curriculum', *Cambridge Journal of Education,* 21(1) (1991), pp. 73–9.

A. GOUGH (1997) 'Education and the environment: policy, trends and the problems of marginalization', *Australian Educational Review,* 39. Canberra: Australian Council for Educational Research.

S. GOUGH, A. W. G. STABLES and W. A. H. SCOTT (2000), 'Beyond O'Riordan: balancing anthropocentrism and ecocentrism', *International Research in Geographical and Environmental Education.*

J. HUCKLE (1993) 'Environmental education and sustainability: a view from critical theory'. In Fien, J. (ed.), *Environmental Education: a Pathway to Sustainability,* Geelong, Deakin University Press, pp. 43–68.

H. R. HUNGERFORD, T. L. VOLK, B. G. DIXON, T. J. MARCINKOWSKI and P. C. ARCHIBALD (1988) 'An environmental education approach to the training of elementary teachers: a teacher education programme'. *International Environmental Education Programme; Environmental Education Series* No. 27, Paris: Unesco–UNEP.

IUCN (1972), 'European Working Conference on Environmental Conservation Education', *IUCN Bulletin* 3(3) (Mar. 1972).

IUCN/UNEP/WWF (1980) *World Conservation Strategy: Living Resources for Sustainable Development*, Nevada: IUCN/UNEP/WWF.

IUCN/UNEP/WWF (1991) *Caring for the Earth: a Strategy for Sustainable Living*, Gland: IUCN/UNEP/WWF.

B. JICKLING (1995) 'Sheep, shepherds or lost' *Environmental Communicator*, 25(6), pp. 12–13.

B. JICKLING (1997) 'If environmental education is to make sense for teachers, we had better rethink how we define it', *Canadian Journal of Environmental Education*, 2, pp. 86–103.

T. J. MARCINKOWSKI, T. L. VOLK and H. R. HUNGERFORD (1990) 'An environmental education approach to the training of middle level teachers: a prototype programme', *International Environmental Education Programme; Environmental Education Series* No. 30, Paris: Unesco–UNEP.

J. MCCORMICK (1989) *The Global Environmental Movement*, London: Belhaven Press.

C. MILBRATH (1984) *Environmentalists: Vanguard for a New Society*, Buffalo, NY, State University of New York Press.

A. MISHRA, U. MALLIK, J. GILL, S. SINHA and D. LAHIRY (1985) '*Environmental education: Pre-service teacher training curriculum development*, unpublished Unesco paper.

NAAEE (1994a) *National Standards for Environmental Education Project Working Paper 1: Introduction to Environmental Educator Performance Standards*, Troy, OH: North American Association for Environmental Education.

NAAEE (1994b) *National Standards for Environmental Education Project Working Paper 3: Developing a Framework for Environmental Educator Performance Standards*, Troy, OH: North American Association for Environmental Education.

A. NAESS (1989) *Ecology, Community and Lifestyle: Outline of an Ecosophy*, trans. Rothenberg D, Cambridge: Cambridge University Press.

T. O'RIORDAN (1981) *Environmentalism*, London: Pion-Methuen.

T. O'RIORDAN (1983) 'The nature of the environmental idea'. In: O'Riordan, T. and Turner, R. K. (eds), *An Annotated Reader in Environmental Planning and Management*, Oxford: Pergamon Press.

T. O'RIORDAN (1989) 'The challenge for environmentalism'. In Peet, R. and Thrift, N. (eds), *New Models in Geography*, London: Unwin Hyman, pp. 77–102.

T. O'RIORDAN (1990) 'On the greening of major projects', *Geographical Journal*, 156(2), pp. 141–8.

C. R. OULTON and W. A. H. SCOTT (1995) 'The "environmentally educated teacher": an exploration of the implications of Unesco–UNEP's ideas for pre-service teacher education programmes', *Environmental Education Research*, 1(2), pp. 213–32.

C. R. OULTON and W. A. H. SCOTT (2000), 'Environmental education: a time for re-visioning in Moon, B., Brown, S. and Ben-Peretz, M. (eds), *The International Encyclopaedic Dictionary of Education.* (London: Routledge).

D. PEPPER (1984), *The Roots of Modern Environmentalism* London: Croom Helm.

QBTR (1993) *Environmental Education: an Agenda for Pre-service Teacher Education in Queensland*, Toowong: Queensland Board of Teacher Registration.

R. E. ROTH (1997) 'A critique of alternatives to national standards for environmental education: process-based quality assessment', *Canadian Journal of Environmental Education*, 2, 28–34.

F. SANDBACH (1980) *Environment, Ideology and Policy*, Oxford: Basil Blackwell.

W. A. H. SCOTT (1994) 'Diversity and opportunity: reflections on environmental education within initial teacher education programmes across the European Union. In F. G. Brinkman and W. A. H. Scott, (eds), *Environmental Education into Initial Teacher Education in Europe (EEITE) 'The State of the Art'*, ATEE Cahiers No. 8, Brussels: Association of Teacher Education in Europe.

W. A. H. SCOTT (1996) 'The environmentally-educating teacher: a synthesis of an implementation theory for pre-service courses', *Australian Journal of Environmental Education*, 12, 53–60.

W. A. H. SOCOTT and C. R. OULTON (1999) 'Environmental Education: arguing the case for multiple approaches', *Educational Studies*, 25(1), 119–25.

S. SELIM (1972) 'Environmental education at the tertiary level for teachers', in *Trends in Environmental Education*, (Brussels: Unesco) pp. 127–44.

G. SESSIONS (1974) 'Anthropocentrism and the environmental crisis', *Humboult Journal of Social Relations*, 2, 71–81.

B. SIMMONS (1996) 'President's message', *Environmental Communicator*, 26(2), 2–3.

B. SIMMONS (1997) *Environmental Education Materials: Guidelines for Excellence*, Troy, Ohio: NAAEE.

W. STAPP, M. CADUTO, L. MANN and P. NOWAK (1980) 'Analysis of pre-service environmental education of teachers in Europe and an instructional model for furthering this education', *Journal of Environmental Education*, 12(1).

S. STERLING (1990) 'Environment, development, education: towards an holistic view'. In J. Abraham, C. Lacey & R. Williams (eds), *Deception, Demonstration & Debate*, Godalming: WWF-UK, & London: Kogan Page.

S. STERLING (1993) Environmental education and sustainability: a view from holistic ethics. In J. Fien (ed.), *Environmental Education: a Pathway to Sustainability*, Geelong: Deakin University Press, pp. 69–98.

D. TILBURY (1992) 'Environmental education within pre-service teacher education: the priority of priorities', *International Journal of Environmental Education and Information*, 11(4), 267–80.

Unesco (1972) *The International Workshop on Environmental Education, Stockholm, October 1972, The Final Report,* Paris: Unesco.

Unesco (1976) 'The Belgrade Charter', *Connect,* 1(1) 1–3.

UNESCO (1977a) *Education and the Challenge of Environmental Problems, Environmental Education, No, 4.* Paris: Unesco.

UNESCO (1977b) *Final Report: Tbilisi,* 14–26 October 1977 Paris: Unesco.

UNESCO (1978) *The Final Report: Intergovernmental Conference on Environmental Education,* Paris: Unesco.

UNESCO (1980) *Environmental Education in the Light of the Tbilisi Conference,* Paris: Unesco.

UNESCO (1992) *State of the Environment Report 1972–1992,* Paris: Unesco.

UNESCO (1977c) *Needs and Priorities in Environmental Education: an International Survey.* ENVED No. 6, Paris: Unesco.

UNESCO (1977d) *Trends in Environmental Education* Belgium: Unesco.

UNESCO–UNEP (1990) 'Environmentally educated teachers: the priority of priorities?' *Connect* 15(1) Mar., 1–3.

UNESCO–UNEP (1992) 'UNCED: United Nations Conference on Environment and Development—The Earth Summit, *Connect,* June, 17(2) 1.

United Nations (1997) *Education and Awareness at Rio +5: Report of the Ad Hoc Committee of the Whole of the Nineteenth Special Session (A/S–19/29; 27 June 1997),* New York: United Nations.

K. WADE (1996) 'EE teacher inservice education: the need for new perspectives', *Journal of Environmental Education,* 27(2) 11–17.

A. E. J. WALS and T. VAN DER LEIJ (1994) 'Alternatives to national standards for environmental education', *Canadian Journal of Environmental Education,* 2, 7–27.

R. WILLIAMS (1985) *Environmental Education and Teacher Education Project 1984–1987,* World-wide Fund for Nature (unpublished).

R. WILLIAMS (1992) *Report on a Survey of Provision for Environmental Education in Initial Teacher Education: Environmental Education and Teacher Education: Preparing for Change and Participation,* Brighton: University of Sussex Education Network for Environment and Development.

WILTSHIRE AGENDA 21 (1997) *Wiltshire Agenda 21 Best Practice Guide,* Trowbridge, UK: Wiltshire County Council.

World Commission on Environment and Development (1987) *Our Common Future,* Oxford: Oxford University Press.

Appendix 1   **Unesco–UNEP Competencies**

Foundational competencies in professional education
The effective environmentally educated teacher should be able to:

- Apply a knowledge of educational philosophy to the selection or development of curricular programmes and strategies to achieve both general education and EE goals. (General education materials and methods may sometimes need merely to be 'environmentalized' to achieve both objectives.)
- Use current theories of moral reasoning in selecting, developing and implementing EE curricula which will effectively achieve EE goals. (Teachers should be competent to use appropriate strategies to allow learners to recognize the role of values in environmental decision-making, clarify value positions and understand the valuing process.)
- Utilize current theories of knowledge/attitude/behaviour relationships in selecting, developing and implementing a balanced curriculum which maximizes the probability of desired environmentally aware behaviour changes in learners. (A balanced curriculum takes into account such aspects as ecological factors vs. trade-off costs, etc.)
- Make use of current theories of learning in selecting, developing and implementing curricular strategies to effectively achieve EE goals. (The methodology of EE as well as the nature of many EE goals is problem-solving. A pragmatic approach on the part of teachers to theories of learning development, such as Piaget's, can do much to increase EE effectiveness in such methodologies and goals as environmental problem solving.)
- Apply the theory of transfer of learning in selecting, developing and implementing curricular materials and strategies to insure that learned knowledge, attitudes and cognitive skills will be transferred to the learner's choices and decision making concerning lifestyle and behaviour. (The ultimate goal of EE is to produce environmentally literate citizens who are willing and capable of taking positive environmental actions in their lifetime.)
- Effectively implement the following methodologies to achieve EE goals: inter-disciplinary, outdoor education, values clarification, games and simulation, case-study approaches, community resource use, autonomous student and/or group investigation, evaluation and action in environmental problem solving, and appropriate teacher behaviours when handling controversial environmental issues.
- Develop and use effective means of planning for instruction.
- Effectively infuse appropriate EE curricula and methods into all disciplines to which the teacher is assigned.
- Effectively evaluate the results of EE curricula and methods in both cognitive and affective domains.

Competencies in environmental education content Level 1: ecological foundations
The effective environmentally educated teacher should be able to:

- apply a knowledge of ecological foundations to the analysis of environmental issues and identify key ecological principles involved;
- apply a knowledge of ecological foundations to predict the ecological consequences of alternative solutions to environmental problems;
- be sufficiently literate in ecology to identify, select and interpret appropriate sources of scientific information in a continuing effort to investigate, evaluate and find solutions for environmental problems;
- communicate and apply in an educational context the major concepts in ecology.

Competencies in environmental education content Level 2: conceptual awareness

The effective environmentally educated teacher should be able to select, develop and implement curricular materials which will make learners aware of:

- how people's cultural or vocational activities (economic, religious, industrial, ect.) affect the environment from an ecological perspective;
- how individual behaviours impact on the environment from the same perspective;
- a wide variety of local, regional, national and international environmental issues and the ecological and cultural implications of these issues;
- the viable alternative solutions available for remediating discrete environmental issues and the ecological and cultural implications of these alternative solutions;
- the need for environmental issue investigation and evaluation as a prerequisite to sound decision-making;
- the roles played by differing human values clarification as an integral part of environmental decision-making;
- the need for responsible citizenship action (persuasion, consumerism, legal action, political action ecomanagement, etc.) in the remediation of environmental concerns.

Competencies in environmental education content Level 3: investigation and evaluation

The effective environmentally educated teacher should be competent to investigate environmental issues and evaluate alternative solutions and to develop, select and implement curricular materials and strategies which will develop similar competencies in learners, including:

- the knowledge and skills needed to identify and investigate issues (using both primary and secondary sources of information and to synthesize the data gathered);
- the ability to:
  — analyse environmental issues and the associated value perspectives with respect to their ecological and cultural implications,
  — identify alternative solutions for discrete issues and the value perspectives associated with these solutions,

— autonomously evaluate alternative solutions and associated value perspectives for discrete environmental issues with respect to their cultural and ecological implications,
— identify and clarify their own value positions related to discrete environmental issues and their associated solutions,
— evaluate, clarify and change their own value positions in the light of new information.

Competencies in environment education content Level 4: environmental action skills

The effective environmentally educated teacher should be competent to take positive environmental action for the purpose of achieving and maintaining a dynamic equilibrium between the quality of life and the quality of the environment (if indeed one can be separated from the other) and develop similar competencies in learners to take individual or group action when appropriate, such as persuasion, consumerism, political action, legal action, ecomanagement or combinations of these categories of action.

Appendix 2   The EEITE Project

Course aims

As a result of pre-service teacher education programmes novice teachers should be both willing and able to make a contribution to environmental education through their own work with learners;

- willing in a sense that they understand the importance of environmental education and have a personal commitment to it which is both practical and intellectual;
- able in a sense that they have a repertoire of management of change and curriculum innovation strategies upon which they can draw in co-operation with others.

These are ambitious aims, and in order to achieve them, pre-service programmes will need to contain two elements. For the sake of clarity these elements are listed here separately. This should not be taken to mean that these will necessarily be separate in practice; rather, tutors will have the responsibility of deciding the inter-relationships between these (and other) elements for themselves, and for determining patterns of organization and support their development work will have. Rather than stifle innovation here, it will be necessary to encourage diversity and to monitor practice in order to gain insights into the transferability of particular approaches and programme designs between institutions.

Programme elements

(1)   aims and practice:

- a consideration of the aims and practice of environmental education, particularly as it relates to compulsory schooling;
- an examination of curriculum practice and extra-curriculum opportunities and the desired learning outcomes associated with these;
- the identification of these characteristics which mark out curriculum activity as contributing to environmental education;
- an exploration of particular strategies and approaches which can be employed in environmental education.

(2)   personal experience in environmental education:

- working with teachers and children in schools on suitably small-scale activities;
- evaluating this practice and building on the foundations laid through reflection and systematic planning;
- in particular, evaluating the effects of this practice on both their own and children's awareness of the possibilities and priorities of environmental education.

It is necessary to emphasize the incremental and iterative nature of such developments, and the consequent necessity of taking a small-step approach, coupled with a focus on the management of intervention and change. The EEITE project evolved a number of didactics characteristics which each institution's developmental project would try to follow. These are such that they might themselves describe desired characteristics of pre-service programmes.

Didactics characteristics

- in part at least, a local focus, drawing from, and contributing to, expertise and awareness in the local community;
- integration in initial teacher education programmes, rather than being an addition;
- a clear set of aims and desired learning outcomes, which are related to the goals of the pre-service programme;
- action-oriented, in that novice teachers will be involved in the planning, implementation and evaluation of the work, and will be encouraged to have an individual commitment to reflection so as to build the experience into their own professional development;
- values and attitude development to be key features;
- processes and outcomes of the work to be able to be shared with other subject didactics groups;
- an interdisciplinary approach, involving more than one subject area or curriculum focus;
- a dual focus, in which tutors and teachers work with novice teachers, who for their part work with students in school.

Appendix 3 The ENSI project

Two basic aims

- To help (school) students develop an understanding of the complex relationships between human beings and their environment through interdisciplinary enquiry.
- To foster a learning process which requires (school) students to develop dynamic instead of static qualities, e.g. exercising initiative, accepting responsibility, taking action to resolve real environmental problems within their locality.

Four guiding principles

- Students should experience the environment as a sphere of personal experience, i.e. by identifying problems and issues within their local environment.
- Students should examine their environment as a subject of interdisciplinary learning and research.
- Students should have opportunities to shape the environment as a sphere of socially important action.
- Students should experience the environment as a challenge for initiative, independence and responsible decision-making.

Theory of learning

- The development of environmental awareness and understanding occurs through an active engagement in finding and implementing solutions to real-life problems that fall within the sphere of students' personal experience.
- Environmental understanding is presumed to develop via a process of systematically reflecting about its application in practice. Learning on the environment is thus viewed as a form of action research.

Appendix 4 Implementation Theory (Scott, 1996)

A necessary outcome of all pre-service teacher education course is that novice teachers should be both willing and able to make a contribution to environmental education through their own work with learners; willing in a sense that they understand the importance of environmental education and have a personal commitment to it which is both practical and intellectual; able in a sense that they have an appropriate repertoire of suitable pedagogical approaches, and management of change and curriculum innovation strategies upon which they, individually and collaboratively, can draw.

In order to achieve this end, pre-service courses need to adopt a teaching and learning process which enables novice teachers, by means of researched cycles of 'practising, investigating', 'systematic evaluation', 'reflection' and 'planning' with others in the school and its community, to develop:

(1) their own understanding of the:
  • aims and practice of environmental education, both locally and globally;
  • desired learning and other outcomes associated with environmental education, particularly, but not exclusively, related to the work of schools;
  • criteria which identify curricular and extra-curricular activity which can contribute to environmental education goals;
  • ways in which they personally might most appropriately, given their own background and interests, contribute to environmental education in schools;

and (2) their own understanding and command of pedagogical and organizational skills associated with developing approaches to teaching and learning which emphasize:

  • interdisciplinary approaches which result in students' understanding of the complex relationships between people and their environment;
  • the centrality of values and attitudes which reflect the goals of environmental education; particularly the need for individuals and groups to confront and appraise their current stances;
  • dynamic ways of working with students so that they develop appropriate personal qualities, e.g. exercising initiative, accepting responsibility, and taking action, which will result in students being able to take co-operative action to address real environmental problems;
  • students' first-hand, active experience in the locality which results in their being able to:
    — identity problems and issues for exploration,
    — research or otherwise investigate such topics,
    — act co-operatively with others,
    — take action to address those problems and issues.

6 The Shaping of French Environmental Policy in the 1990s

Joseph Szarka

Summary

In France the 1990s have been a period of renewed activity in the production of environmental policy and legislation. The strengthening of environmental policy has been favoured by a combination of social and political factors. The relative lull in environmental policy-making during the 1980s fostered the necessity for France in the 1990s to catch up with international developments and to implement European initiatives. Renewed social demand for environmental protection translated into electoral pressure at the ballot box and a more favourable climate for green policies. This in turn provided openings for ecology parties and opportunities for action from entrepreneurial Ministers of the Environment. Institutional capacity was enhanced by an increase in resources and administrative reorganisation. Finally, a broader range of instruments was developed, going beyond the production of regulations and placing more stress on fiscal and market-based measures. This chapter reviews and assesses these developments and suggests directions for future research.

I INTRODUCTION

Not only have environmental issues developed greater salience in their own right internationally but also their differential uptake across countries provides one indicator of the pace of political renewal at the level of the nation. Clearly, national polities can be more or less receptive to new agendas. Further, study of environmental policy streams provides fresh insights into the functioning of national political systems because the transversal challenges posed by environmental

116

issues create new demands (in terms of institutional capacity and co-ordination) and highlight social and economic conflicts (related to priority setting, resource allocation, etc.). This produces a two-way research agenda: study of environmental policy illustrates the capabilities and limits of a national polity while study of a national polity reveals the opportunities and constraints for environmental policy-making.

Although the nation state is not the sole actor in these processes, it remains as one central protagonist. Consequently, nation based studies of environmental policy occupy a major place in the literature, since policy initiatives still require national impetus and backing (and downstream commitment), regardless of whether the sponsor be governments, lobbies, NGOs or public opinion, or whether the arena for discussion is primarily domestic, European or international. However, the quantity and quality of such studies vary consequently, with considerable attention devoted to the USA, UK and Germany, but less to other countries. The more limited coverage of France, for example, is surprising, as the country remains a major power and one of the leading world economies, with a distinctive and rich political culture, a highly organized administrative system, and one of the largest and most ecologically varied territories in Europe.

While a reputation for being a follower rather than a leader in environmental policy provides some explanation for this situation,[1] this interpretation calls for qualification and review. In France, the 1990s have been a period of renewed activity and major reform in the production of environmental policy and legislation. The strengthening of environmental policy has been favoured by a combination of social and political factors. The relative lull in environmental policy-making during the 1980s fostered the necessity for France in the 1990s to catch up with international developments and to implement European initiatives. Renewed social demand for environmental protection translated into electoral pressure at the ballot box and a more favourable climate for green policies. This in turn provided openings for ecology parties and opportunities for action from entrepreneurial Ministers of the Environment. Institutional capacity was enhanced by an increase in resources and administrative reorganization. Finally, a broader range of instruments was developed, going beyond the production of regulations and placing more stress on fiscal and market-based measures. This article will explore each of these shaping factors in the development of environmental policy and make a critical evaluation of progress made to the end of 1997.

II   SOCIAL AND POLITICAL INFLUENCES

The end of the 1980s was a period of increased concern over environmental issues. Public opinion was mobilized by heavy media coverage. With 1989 as the 'Year of the Environment', European citizens were regularly confronted by issues such as acid rain and 'global warming', which taken with the hot summers of 1989 and 1990, seemed to give support to the view that environmental issues had changed in scale. In France, increased public concern translated into a significant green vote. From being a motley and marginal collection of interest groups in the 1970s, French ecologists gradually developed political organizations, identity and credibility. The *Verts* came into being as a political party in 1984. In the 1989 European elections, despite a radical programme, they polled nearly 2 million votes (10.6 of votes cast). Proportional representation allowed them to return nine MEPs to Strasbourg. In 1990, the Minister for the Environment, Brice Lalonde, formed a rival party, *Génération Ecologie*. Lalonde had been a militant in the new Left in the early 1970s, chairman of French Friends of the Earth and had contested a number of elections, including the 1981 presidential election. In the 1992 regional elections, *Génération Ecologie* campaigned on a more reformist or 'realist' programme than the *Verts*, and attracted 7% of the vote as compared with 6.7% for the latter. The aggregate green poll of 3.5 million voters proved to be an electoral high.

In elections fought between 1993 and 1997, the turnout for the ecologists has been much lower. In the 1993 parliamentary elections, the *Verts–Génération Ecologie* coalition failed to win any seats. However, under a first past the post system and with a landslide victory by the right, the poll of 7.6% remained creditable. Yet the ecologists were bitterly disappointed. The ensuing in-fighting prevented a coalition at the 1994 European elections, with the result that two separate ecology lists were put forward of which neither could pass the threshold of 5% of the poll required to send representatives to the European parliament. Again the ecologists won no seats. The in-fighting continued in the 1995 presidential election, when Lalonde, Waechter[2] and a new leading light of the *Verts*, Dominique Voynet, all proposed to stand in the election. Only Voynet was able to do so. However, in a context where the electorate were more concerned with employment than the environment, her campaign failed to spark interest or retain the green vote from previous elections, leading to a low outturn of 3.3%. The French greens seemed buried, if not quite dead.

Yet in the snap parliamentary elections of 1997, they reached out from the grave and won eight seats in the *Assemblée nationale*. This unexpected outcome was achieved on the back of a broad left alliance led by the Socialist Party. Knowing that the Right was unpopular but that the elections were too close to call, the strategy of the Socialists was to garner the widest vote possible. They agreed to stand down in certain constituencies and let their green allies through to the second round. Moreover, the electoral pact extended to the distribution of ministerial portfolios. The new Jospin government included Communists and a *Vert*, Voynet, as Environment Minister.

The roller-coaster fortunes of French political ecologists lend support to the thesis of an 'issue-attention cycle' (Downs, 1972), in which public interest in the environment periodically waxes and wanes. But they also demonstrate the perennial nature of the green movement. Far from being a marginal phenomenon that can safely be ignored, French green parties have been able to mobilize millions of voters. Because the French political system is relatively open and fluid (by contrast with Britain's more closed polity), green politicians have been able to avail themselves of specific opportunity structures that lever up their political impact. Thus they have participated at almost every level of the polity—with mayors, representatives in the councils for the *départements* and the regions, and now members of parliament. Even presidential elections, though impossible to win, have furthered the green cause by providing a national forum at modest cost and by providing a trampoline to ministerial office. Two green presidential candidates—Lalonde and Voynet—have gone on to be appointed as Environment Minister in coalition governments.

These developments give the lie to a number of received ideas. First, 'greenery' is not a passing fashion. Secondly, although the green movement in France in the 1970s and 1980s was considered weak by comparison with countries such as Germany, Denmark and the Netherlands, over time it has made a comparable impact. Thirdly, contrary to the complaints of many greens, the French political system does not automatically exclude small parties from political life, or even from the legislature and the executive. In the 1990s, French political ecologists contributed directly to the formulation of environmental policies to an extent that remains unthinkable within the British or American systems. This does, however, raise the issue of performance in office and imposes the strain of realism and responsibility on the utopic currents ever present within the green social movement.

III   MINISTERIAL ENTREPRENEURSHIP

French Environment Ministers have suffered from a number of handicaps: limited resources, a tiny administration, lack of political clout and the suspicion that the environment is indeed their top priority. Making headway under these circumstances has been slow and difficult, but the gradual increase in salience of environmental issues provided the pre-conditions for major initiatives in recent years.

By the start of the 1990s, the need to upgrade French policy to international levels was perceived by politicians and government. Following countries such as Canada and Holland, France put together a 'green plan'.[3] Developed by Environment Minister Lalonde and adopted by the Rocard government on the 19 December 1990, it recommended a paradigm change in environmental policy. The plan listed the shortcomings of French environmental policy over the previous 20 years as unambitious targets (especially related to water quality and noise), excessive State centralisation, a weak Environment Ministry and a narrowly French (rather than international) view of environmental policy-making. Consequently, France had exercised little influence in the EC and other international agencies, and had not been involved in the Brundtland report. Meanwhile the scale of hazards had increased, good environmental performance had become a competitive advantage and a clean environment was seen as a right. The plan's national environmental audit detected no catastrophic problems but the overall situation was considered unsatisfactory. The position was reported as 'good' in relation to water and energy consumption, available natural space, treatment of toxic waste, cleanliness of beaches and exposure to acid rain; as 'middling' with respect to levels of fertiliser and pesticide usage, and treatment of household waste, but poor as regards transport pollution, noise levels, numbers of waterworks and defences against forest fires. The nuclear industry was seen as presenting particular difficulties, notably regarding the disposal of radioactive waste.

Remedial proposals included greater decentralization, more co-operation between the State and local authorities, experiments in greater democratic participation and public involvement in decision-making (with a bigger role for environmental groups), more solidarity and a more international outlook. Lalonde's green social movement background is evident here, as it is in his choice of an ambitious level of environmental protection. In terms of major reforms, he proposed a restructuring of the Ministry of the Environment and its field services to increase its effectiveness (discussed below). He called for the establishment of a specialized corps of environmental inspectors, in order to overcome the lack of specialist expertise within the ministry. Legal reform was proposed in the form of a *Code de l'environnement* to consolidate France's disparate range of legislative provisions related to the environment.

Hatem (1991) criticized the plan for its lack of clear priorities, evasiveness about costs and excessive stress on the role of the State. However, the water and waste acts passed during Lalonde's period in office, as well as the reorganization of environmental protection agencies, represented significant progress. The underlying difficulty in implementation was political. Lalonde outlined to 10-year time horizon for his policy; this long-term view was appropriate in environmental terms but political deadlines were far shorter. With the popularity of the Socialists plummeting in 1992, Lalonde abandoned the sinking ship and resigned, giving himself the scope to contest the parliamentary elections of 1993 on a green ticket.

His successors at the Environment Ministry have not lacked ambition but have been short on resources and enjoyed even less time to see their reforms through. Ségolène Royal's tenure in office was barely a year. Her ambitions can be measured by a speech given at Rio, in which she sketched out a massive project of social and environmental renewal, promising French support for initiatives in combating the enhanced greenhouse effect and protecting biodiversity and forests, together with increased development aid (Royal, 1994). Practical measures were modest, though occasionally spectacular, such as unilateral halting of imports of household waste from Germany—despite the lack of a clear legal basis for this action in EC legislation. Subsequently, the Council of Ministers declared rigorous restrictions on both the import and export of waste in the EC, thereby giving some priority to environmental concerns above the free movement of waste 'goods'.

In March 1993, after the massive victory in the parliamentary elections of the right-wing RPR–UDF coalition, Michel Barnier was appointed Environment Minister. Barnier's initiatives had innovative elements but proved rather piecemeal. He attacked on a number of fronts including another reorganization of the Environment Ministry and developing publicity campaigns to raise environmental awareness among the general public. The 'first national conference' on the environment was launched, with Prime Minister Balladur as keynote speaker. A major national debate on energy sourcing and utilization was organised in 1994 in collaboration with the Ministers of Industry and of Higher Education and Research. Thus on the 20th anniversary of the launch of France's massive nuclear energy programme, a public debate was finally held on the relative merits of nuclear energy versus fossil fuels, renewable energy sources and energy conservation. In the event, the French public were unenthused by the debate: the national 'consensus' on nuclear based electricity remained unruffled. In the follow up to the Rio Summit, a *Commission française du développement durable* was set up, meeting for the first time on 12 April 1994. France stepped up aid to developing countries in the fields of water resources, use of tropical forests and energy utilisation. The extent of his environmental ambitions can be gauged from the contents of the *loi Barnier* of 1995 (discussed below), but—as with Ségolène Royal—electoral developments curtailed his period in office.

In the major government reshuffle subsequent to Jacques Chirac's victory in the 1995 Presidential elections, Barnier left the Environment Ministry to become Minister for European Affairs and Corinne Lepage was appointed to the Environment portfolio. Unlike her immediate predecessors, Lepage was not a career politician, but a prominent environmental lawyer who favoured a tougher stance on environmental protection. Logically enough, her period of tenure was marked by vigorous objectives for the upgrading of legislation, notably by an air pollution statute and work on codifying French environmental legislation (discussed below). However, her lack of political experience and limited support from Prime Minister Juppé limited her effectiveness and scope for innovation. This is a crucial factor, in that in the French system the Prime Minister exercises considerable constitutional powers of decision-making and arbitration in interministerial disagreement. In March 1997, when Juppé announced the restarting of the Superphénix nuclear fuel reprocessing plant in Creys-Malville, French greens organized street protests and Lepage joined in the criticism, apparently embracing a strategy of 'double or quit' within a government and parliament showing increasing signs of indifference to environmental issues.

In the event, the surprise defeat of the right in May 1997 opened the way for Dominique Voynet, the principle spokeswoman of the *Verts*, to be appointed as Minister for Regional Planning and the Environment. This enlarged portfolio gave her more influence in government than her predecessors and so greater capacity to formulate effective environmental policy. Unlike Lepage, Voynet had political experience and capital, but with only eight green *députés* in Parliament opportunities to exert direct influence were limited. Strategies of co-operation and persuasion within the government team were of the essence. She did, however, benefit from the support of Prime Minister Jospin. In June 1997, Voynet was able to announce two highly symbolic measures: the halt of the Superphénix nuclear plant and the cancellation of the Rhine–Rhône canal. Both of these two schemes had for years attracted heavy criticism from French greens as environmentally damaging and unnecessary. At the time of writing, the French social and political context seemed ripe for an ambitious environmental policy. Success will depend on the extent to which Voynet seizes opportunities for renewal at national and international levels. Of particular interest will be responses to sustainable development and local Agenda 21s, which are still in their infancy.[4]

IV   EUROPEAN INFLUENCES

EC/EU legislation has been instrumental in pushing out the boundaries of French environmental policy and filling in a series of normative objectives related to environmental quality. As pointed out by the *Institut français de l'environnement* (1994), environmental policy in the 1990s has often been a catching up process, with the transposition and application of a significant number of European directives. Examples include: the 1992 Water Act; the 1993 Landscape Act (which brought environmental considerations within the purview of building planning regulations); the 1992 Household and Industrial Waste Act; the 1992 Noise Abatement Act; the 1993 Public Inquiries and Environmental Assessment Act (which finally transposed the 1985 European directive regarding public information and evaluation of cross-border impacts); the 1993 Quarries Act (placing quarries into the category of 'scheduled works' and henceforth regulated by the Ministry of the Environment, not the Ministry of Industry); the 1992 Biotechnologies Act; the 1995 *loi Barnier* (containing a number of broad principles and cross-cutting measures) and the 1996 Atmospheric Pollution Act. Clearly, within this policy stream, technical and normative orientations predominate. Indeed, commentators such as Mény (1995) and Majone (1996) have described the EU as predominantly a 'regulatory state', an observation which has considerable relevance for its environmental regime. However, a list of directives and laws gives few clues about the political and social context from which they emerge or the degree of success with which they are implemented. Here two broad questions emerge—why have nation states such as France subscribed to this regulatory framework and how has it been implemented?

Integration of EC/EU environmental policy has been only one chapter in a wider drive towards European integration. France's European strategy has formed part of a grand design since the 1950s, with the consistent pursuit of national goals by European means and political ambitions by economic methods.[5] The idea of *une Europe à la française* marked not only the policy of General de Gaulle but also his heirs. However, environmental policy has not figured prominently in the grand design. In the 1980s, France seemed satisfied to accompany environmental initiatives by Germany, Denmark or the Netherlands who have taken a more pro-active stance. France's concern with the 'high politics' of integration has resulted in a measure of attention displacement away from the 'low politics' of environmental regulation. The latter has largely been treated as a contingent instance of 'spillover' politics, an example of day-to-day negotiation apparently disconnected from 'history-making' decisions, such as the Maastricht Treaty. Yet this low profile approach should not be interpreted as outright reluctance. In its survey of French environmental policy, the OECD (1997, p. 211) indicated that, as at 1995, France had transposed almost all European environmental directives into national legislation and had notified Brussels that it had done so in relation to 118 directives out of 125.

In response to the question of why governments sign up to stringent environmental policy, Golub (1996: p. 21) put forward the hypothesis of 'slack cutting': 'environmental ministers enjoy interministerial slack when cutting deals in Brussels which have been blocked domestically by traditionally powerful ministries such as Finance, Trade and Industry, or Agriculture'. In consequence, tighter regulations than their domestic ministerial colleagues would have countenanced are brought in via the EU Council of Ministers, in those sessions where only Environment Ministers deliberate. Further, Collier and Golub (1997) suggested that issue linkage (where negotiations lead to trade-offs), the constraints of intergovernmentalism (veto is not always an option) and expected non-compliance (the belief that targets are to be aimed at rather than hit) are further reasons explaining willingness to sign. These observations point up the in-built tensions that characterize environmental policy: the tendency to make promises that cannot easily be kept may jar certain sensibilities, yet without setting higher standards it is doubtful that performance will ever be improved.

Inquiry into the manner of implementing European legislation is also instructive. Unlike the British case, where EC/EU legislation is often highlighted as alien, in France's case European directives undergo a process of 'naturalization', with the result that the French are often unaware of the influence of EC/EU measures on national legislation. Explanations lie in cultural and procedural factors. First, France as a founder EEC member still feels that it has a leading role in setting the rules and agenda for the European 'club', and that the benefits far outweigh the drawbacks. Thus scapegoating of Brussels bureaucrats rarely makes sense. Secondly, the Community origin of much domestic legislation is rendered almost invisible by the nature of transposition procedures. In France a choice exists between parliamentary or ministerial routes. During the Fifth Republic, the powers of the parliament have been limited in comparison with earlier periods of French history or nations such as Britain. Article 34 of the constitution identifies those areas in which Parliament can legislate by statute (*loi*), whereas article 37 specifies that in all other matters the executive legislates by regulation (*règlement*). European integration has tended to sideline Parliament further in that it has no direct input into the formulation of European legislation. In a parliamentary report, Ligot (1991, p. 20) noted that of some 1,000 European directives in effect in 1991, only 75 had been transposed by statute. Thus most directives, including environmental ones, are transposed by ministers (Simon and Rigaux, 1991, p. 293). As a result, there is less scope for debate and public scrutiny. Thirdly, European legalisation is usually incorporated into French administrative law (a branch of law with no discrete existence in Britain), which conditions the behaviour of public sector bodies. This too lessens its visibility—and even its value—for the general public.

Discretion reduces controversy, but can also backfire. For example when, shortly prior to the passing of the 1991 'nitrates' directive (which aimed to reduce levels of nitrates leaching from agricultural land into water supplies), Environment Minister Lalonde pointed out that French farmers were responsible for water pollution in rural areas, the butt of the anger of the agricultural lobby was not the EC directive but the French minister. In this over-heated climate, it took several years for the statement to be accepted as valid and for remedial action to be initiated. However, this case also reinforces the point that one route to promote contentious legislation is via Brussels, allowing ministers and officials to defuse tension by evoking the desirability of being a 'good European'—a strategy that remains legitimate and profitable in French political culture.

V NEW POLICY AND LEGISLATION

During the French presidency of the EC in 1989, Lalonde noted at the Council of Ministers that some 80% of environmental directives concerned pollution, discussions were of a highly technical nature and so broader questions relating to land use, nature protection and citizens rights were hardly debated (Auer and Massiet du Bierst, 1990, p. 167). Given the constraints related to transposition of EC directives, much French legislation in the 1990s likewise exhibited these characteristics. However, significant attempts were made to go beyond narrow, technical formulations.

The major example is the *loi Barnier*, which went on the French statute books on the 2 February 1995 after 18 months of parliamentary debate and over 1,000 amendments. It is the largest and most ambitious piece of French environmental legislation to date. Barnier (1996: p. 200) described it as *'le premier texte transversal en matière d'environnement'*. By way of actual measures, the bill provided for greater public participation in enquiries related to large-scale state projects (such as motorways, high-speed railways, etc.). A new *Commission du débat public* was created to ensure that a wider range of views was considered in the planning stage. The *loi Bourchardeau* of 1983 on public inquiries was revised and updated—a decade of experience had demonstrated that French public inquiries were largely token gestures. Environmental groups were given new rights to bring court actions in relation to infringements of laws on quarries, landscapes and radioactive waste. Regarding natural catastrophes, new encouragement was given to local authorities to draw up prevention plans, by simplifying procedures and making more resources available to communities at risk. Expropriation procedures for the most dangerous areas were simplified. Nature conservation measures included a national inventory of sites, as well as increased prerogatives and resources for national and regional parks. Charges for landfill waste disposal were increased and a new tax was introduced on 'special' industrial wastes to finance the clean-up of abandoned tips.

But it was its announcement of principles which aroused most debate. The bill set out a *droit à l'environnement* (a general right to a safe environment)—an innovation which had been under discussion for many years. The new right was originally envisaged by Barnier (and others) as an amendment to the Constitution: this was abandoned as too ambitious a proposal, thereby diluting its import. In addition, the bill stressed the polluter-pays principle, prevention at source, the precautionary principle, progress towards sustainable development and the public's right to information and participation. These principles were less novel, being well-known in the environmental literature and present in a number of intergovernmental agreements at European and international levels. Most figure in both the Single European Act of 1987 and the Maastricht Treaty of 1992, so in effect France already subscribed to them.[6] The transposition into national legislation was the opportunity for parliamentarians and lawyers to explore how legal force was to be given to these general and abstract concepts. At one level, the large number of amendments reflected the need to clarify the bill's content and facilitate its implementation, but they also point to the scale of the legal difficulties raised. Yet despite the lengthy debate, the precise consequences were not worked out. As Cans (1995, p. 208) noted, critics feared little of substance was being incorporated into law.

This stress on general principles can be better understood by reference to the context in which the law was drafted. On the one hand, French environmental legislation was limited to a number of piecemeal measures, often enacted by government decrees rather than by statute, which regulated specific media (water, air, soil) or domains of environmental concern (nature protection, land use, etc.). This legislation was extremely fragmented and consigned under various pre-existing categories of law (town-planning, industrial pollution, countryside law, etc.). Moreover it contained few over-arching statements of aims to guide the courts in their application of particular, usually highly technical, measures. Environmental legislation was largely limited to being a sub-branch of administrative law (governing the actions of public bodies), rather than giving individuals and groups specific rights to defend environmental objectives. A number of negative consequences resulted from this state of affairs. Bringing cases before the courts in instances of environmental damage was arduous or even impossible. The limitations of legal texts made the decisions of the courts difficult to call. Rather than acting as a deterrent on polluters and destroyers of the environment, the legislation tended to dissuade the victims of damage from taken legal action. Moreover, recourse to the courts usually presupposed that harm had been done and offered a means of redress and repair. But when forests are flattened, marsh-lands drained, natural resources polluted or species annihilated, the loss can be irreversible. Although nature conservation will always be vulnerable to the *fait accompli*, there was a clear need to reinforce the logic of repair (where repair was possible) and to introduce a logic of prevention (especially where potential loss was irreversible). However, the shift from a technical and normative environmental regime to a framework of environmental rights and principles—or more pointedly, from a coercitive to an emancipatory ideology—has proved more complex than expected.

The scattered nature of the environmental legislation was particularly flagrant in that many branches of French law have been rationalised in the form of a 'code'—a single volume of law where measures related to a particular domain (such as commerce, labour law, countryside law, etc.) are systematically presented. This makes comprehension and application of the law more predictable and 'safe'. A call to consolidate French environmental legislation by a *Code de l'environnement* was already present in Lalonde's 'green plan'. The *loi Barnier* paved the way for parliamentary debate on a *Code de l'environnement* and it fell to Lepage, as Environment Minister and experienced lawyer to translate this intention into a parliamentary bill in 1996. However, its progress was delayed by much-publicized snipping from parliamentarians over minor technical points and even spelling errors. The early dissolution of parliament in April 1997 interrupted its passage, and at the time of writing the bill remained pending. Overall the episode illustrated the difficulty of greening a conservative government and parliament, especially where the environment minister has little political clout.

With air pollution in major cities reaching critical levels, Lepage's other major initiative was to bring a new act on air quality before parliament in June 1996. The act contained measures for more accurate monitoring of air pollution levels in cities and instituted a graduated series of alerts to inform the population of excessive pollution levels. Preventive action in the form of transport strategies was encouraged and local authorities were given powers to take exceptional action. These included *circulation alternée*, whereby only 50% of vehicles for personal use are allowed to use the roads on a particular day. The measure was applied for the first time on 1 October 1997 throughout

Paris and its conurbation, and public transport was also made free. The bill offered a salutary reminder that the consequences of significant environmental initiatives tend to be unpredictable. Despite fears of widespread flouting, it proved a considerable success with only 5% of motorists in contravention. Opinion polls showed that over 80% of greater Paris residents agreed with the measure. Public opinion had been well-prepared for this eventuality. It was originally floated by Lepage in press interviews during July 1995. The bill's passage though Parliament in 1996 continued to publicize the issue, while repeated occurrences of poor air quality in the summer of 1997 led commentators to criticize Voynet for not imposing available restrictions sooner. Thus implementation was widely expected. Interestingly, a bill characterized by a technical bent and a crude 'command and control' ideology demonstrated that one of the keys to successful policy implementation is the generation of public goodwill. Though the bill had been widely criticized for lacking teeth, it provided the most spectacular environmental measure taken to date in France. This is not, of course, to say that the problems of traffic congestion were solved. Rather stress is placed on the difficulties encountered by policymakers in predicting and evaluating the effects of environment policy at planning and decision-making stages.

VI INSTITUTIONAL CAPACITY BUILDING

Even within the administrative domain, improved competence has required trial and error and lengthy battles. Institutional capacity-building has been an essential prerequisite and accompaniment to implementation of environmental legislation. The French Environment Ministry was set up in January 1971, but its remit and resources were limited and subject to repeated challenges from vested interests in powerful ministries such as Industry and Agriculture. Carving out a niche for itself within the previously established networks of power in French government has proved a slow process. By the 1990s, the Environment Ministry had learnt to play a two-fold role: direct involvement in certain sectors and co-ordination of other ministries on environmental matters.

Hands-on action has required building up agencies and field services which, at the Environment Ministry's inception, either did not exist at all or were attached to other ministries. Ambiguities over remits of over-lapping or competing administrations resulted in extreme fragmentation and limited effectiveness.[7] In 1990, under Environment Minister Lalonde's 'green plan', the regional field services of the Industry Ministry were reorganized, given additional responsibilities for the environment and re-labelled as the DRIRE (*Directions régionales de l'industrie et de la recherche et de l'environnement*). The DRIRE report on environmental matters to the Ministry of the Environment. They co-ordinate the *Inspection des Installations Classées* (Inspectorate for Scheduled Works) and enforce regulations on industry related to pollution and waste. At the same time, Lalonde established the *Directions régionales de l'environnement* (DIREN), to act as the regional representative of the Environment Ministry in relation to water, architecture and town planning, and nature conservation. These developments enhanced the institutional means to ensure compliance with the widening raft of European and national legislation. However, reform was limited by the resistance of other ministries and by the inbuilt complexities of managing environmental resources.

Contrary to the reputation of the French for centralized intervention and despite the apparent rigidity of environmental limit values and quality objectives, enforcement by the regulatory agencies (the DRIRE) has generally been based on negotiation, rather than on heavy-handed coercion.[8] Compromises are reached between divergent and competing interests, principally: firms preoccupied by their costs base, environmental groups, the general public and the local authorities, who are concerned with keeping the peace and maintaining continuity of political power. The DRIRE are bound by the administrative and political processes in which they operate. Their objectives are both to promote industrial development and limit pollution: though not automatically incompatible, these two targets can pull in different directions.[9] At the local level, the DRIRE are responsible to the *préfet*[10] and it is the latter who, by virtue of the 1976 statute on classified establishments, authorizes pollution discharges and specifies operating procedures in line with environmental and safety standards. Thus the problems of industrial pollution are conceptualized primarily in terms of compliance with or infringements of administrative permits, rather than in terms of environmental damage. However, in conditions of economic exigency, industrialists are prone to pressure politicians, who in turn lean on the *préfet*. Accommodation to these conflicting demands has been the norm, which can result in inspectors tolerating a level of non-conformity with environmental regulations. Outright flouting or negligence seems to be rare; rather non-compliance is conditionally accepted so long as investments in improved technology come into play within deadlines agreed between industrialists and inspectors. But deadlines can slip. The major outbreak of fire at the Protex factory in 1988 exemplified the dysfunctional nature of this system.[11] Legal proceedings instigated by inspectors are very much a last resort, as they are expensive and uncertain in their outcome, with administrative pressure considered to be more effective. But the balance of influence between administrators and politicians at local and regional levels proves problematic. Elected representatives typically seek to take credit for environmental gains for themselves, while passing the dirty work of clamping down on polluters back to administrators, leaving the latter with the recurrent dilemma of how to arbitrate between environmental and economic concerns. Because the environment presents a new, highly charged arena for negotiations capable of generating significant political gains and losses, more effective enforcement requires both greater resources (especially a larger number of inspectors) but also greater *political* backing for enforcement agencies.

VII ECONOMIC AND MARKET BASED MEASURES

Although regulation has been the norm for environmental policy in France, its limitations have been progressively recognized. Effective enforcement requires considerable administrative and political resources. In conditions of rapid technological change and growing complexity of environmental problems, making and enforcing detailed regulations has proved increasingly problematic. In the late 1980s, in the context of deregulation and broad ideological change stressing markets rather than the State, direct intervention lost favour whilst the use of market instruments grew more attractive. With the latter, the aim is to incorporate the impact of externalities, rather than leave them uncosted until cumulative damage mounts to excessive levels, which for too long has been received practice.

An early instance of the application of levies is provided by the *Agences de l'eau* (water boards) which have used them since the 1960s to finance water treatment costs, in application of the polluter pays principle. A more recent innovation was the establishment of the firm Eco-Emballage (on the 1 January 1993) to promote the reuse and recycling of packaging; its activities are financed by charges. However the key example is that of the *Agence de l'environnement et de la maîtrise de l'énergie* (ADEME)—the Environment and Energy Saving Agency—which was established under Lalonde by act of parliament on the 19 December 1990. It brought together three previously separate agencies for air quality, waste and energy, though the process of amalgamation was to prove troublesome for several years, especially as the agency is responsible to no less than three ministries (the Ministry of the Environment, the Ministry of Industry and the Ministry of Higher Education and Research). In 1993, the ADEME employed approximately 600 staff and had an annual budget of 1.2 billion francs. It was enjoined the task of collecting a number of environmental levies and charges, based generally on the polluter pays principle. These included a charge on waste disposed by landfill, a tax on oil products to finance collection of used oil, a tax on air pollution from fixed sources and a tax on airport noise. Revenues are redistributed in the form of research grants, subsidies and loans to firms to encourage clean technologies and improve environmental performance.

A frequent criticism of these levies has been that they tended to be set too low to change behaviour: they have contributed to financing the costs of pollution clean-ups, rather than averting environmental damage. In response, increases in levies were implemented in the mid-1990s, with water pollution levies doubling over 1992–6 and landfill disposal costs rising from 20 francs per tonne in 1992 to 40 francs in 1998. As charges mount, the deterrence effect increases, but a new question arises namely, the payment level at which illegal dumping becomes irresistible (at least for the unscrupulous), with disproportionate costs for the community. As the intention is to eliminate landfill disposal altogether by 2002 (other than for 'final' wastes) and with fly-tipping already rife, the viability of official policy has been called into doubt.

In the fiscal field, the most ambitious suggestion in recent years has been the call for a carbon dioxide tax, with the intention of counteracting global climate change. The European Community's proposal for such a tax was opposed by the French, on the grounds that it would disadvantage the French electro-nuclear industry (which is the largest in Europe and produces negligible CO_2) and that its actual effects had not been properly assessed, particularly as regards industrial competitiveness on world markets.[12] The example illustrates the contention that the precautionary principle *is* applied by policy-makers, but not necessarily in the manners in which environmentalists would hope.

VIII CONCLUSIONS

Two key themes have run through this analysis of the shaping factors on French environmental policy. One is that policy became bolder and more vigorous in the 1990s. The other is that policy was marked by uncertainties over best available means (in the broad sense) and limited ability to predict outcomes.

Policy-makers improved their competence, but largely by trial and error. The limits on their knowledge due to the paucity of environmental information were acknowledged and the task of data gathering was enjoined to the *Institut français de l'environnement*, which is France's major interlocutor with the European Environment Agency. This mode of data treatment remains orientated towards scientific and technical information. While the value of this species of knowledge should not be underestimated—particularly if it highlights environmental protection shortcomings and exposes major polluters—it is insufficient in itself. In the environmental domain, questions of issue framing, policy formulation, implementation, co-ordination and evaluation require a comprehensive and transversal understanding of political and social processes. A call for significantly more research in these areas will come as no surprise. What is surprising is that far less research has been conducted in France than in Britain even on fairly obvious topics such as the role of public opinion and environmental groups in policy development, the effectiveness of enforcement agencies, or the environmental benefits of market-based instruments.

The accumulation and heightened visibility of environmental problems, together with the universal aspiration to a healthy lifestyle and rising standards of well-being, pose challenges that policy-makers no longer ignore. At the same time, the nature, direction and effectiveness of environmental policies are conditioned by their social and political context. Rather than being simply a collection of technical measures to combat the side-effects of technological progress, environmental issues have deep roots in social and economic practices. The ways in which problems are framed and addressed are themselves the products of those social and cultural processes which have traditionally been the subject of sociological study. Further, national and international political systems are prone to change. This article has shown that the characterization of the French polity as 'closed' to green views which prevailed among political ecologists was overstated. In the late 1980s and 1990s, strategy changes effected by political parties enabled more effective utilisation of the political opportunity structure. This in turn opened the way for new environmental policy initiatives. To indicate that there have been limits on the scope of those initiatives is to state the obvious. Of more interest to the researcher and practitioner is their illustration of the difficulty of predicting the outcomes of policy processes. While it will never be the business of the academic researcher to gaze into the crystal ball, limitations on the ability to link cause and effect in contemporary policy arenas such as the environment raise a worthwhile challenge for the political sciences.

Notes

1. For comparative studies of national propensities to take environmental initiatives at the EC/EU level, see Sbragia (1996) and Collier and Golub (1997).
2. Lalonde and Waechter had both previously contested presidential elections, in 1981 and 1988 respectively.
3. For a cross-national study of 'green planning', see Dalal-Clayton *et al.* (1996). For a presentation of France's green plan, see Theys and Chabason (1991).
4. Progress reports on Rio commitments can be found in Commission française du développement durable (1996) and Comité 21 (1996).
5. See Duchêne (1996) and Moreau Defarges (1996).
6. Curiously the *loi Barnier* made no reference to the 'integration principle' which is present in the Maastricht Treaty (whereby the integration of environmental concerns into other policy sectors is a legal requirement).
7. Romi (1990, pp. 32–45) gives detailed examples of these problems.
8. This point is repeatedly made in the literature. See for example Lascoumes (1994, p. 153) and Dron (1995, p. 66).
9. Recommendations for a separation of enforcement from consultancy and advice services have yet to be implemented.
10. Charbonneau (1994) provides a telling discussion of the administrative and political tangle this generates.
11. A study of the Protex case can be found in Lascoumes (1994, pp. 95–6).
12. The French Senate report prepared by François (1996) is revealing in these regards.

Further Reading

M. BOGIDUEL and H. BULLER (1995) 'Environmental Policy and the regions in France', in Loughlin, J. and Mazey, S. (eds) *The End of the French Unitary State? Ten Years of Regionalisation in France (1982–1992)*, pp. 92–109, London: Frank Cass.
C. LARRUE and L. CHABASON (1998) 'France: fragmented policy and consensual implementation', in Hanf, K. and Jansen, A.-I. (eds) *Governance*

and Environment in Western Europe. Politics, Policy and Administration, pp. 60–81, Harlow: Longman.

OECD (1997) *Environmental Performance Reviews: France*, Paris: OECD.

J. SZARKA (forthcoming) *Environmental Policy in France. The Shaping of a New Agenda*, Oxford: Berghahn.

References

L. AUER B. and MASSIET DU BIERST (1990) 'La France et la politique européenne d'environnement', In ENGREF (ed.) *La Politique européenne de l'environnement*, pp. 166–74, Paris: Editions Romillat.

M. BARNIER (1996) 'L'apport du loi', *Revue française de droit administratif*, 12:2 (mars-avril), 200–2.

C. CANS (1995) 'Grande et petite histoire des principes généraux du droit de l'environnement dans la loi du 2 février 1995', *Revue juridique de l'environnement*, 2, 195–217.

S. CHARBONNEAU (1994) 'Administration du travail et administration de l'environnement', in Société française pour le droit de l'environnement (eds) *Droit du travail et droit de l'environnement*, pp. 127–32, Paris: Litec.

U. COLLIER and J. GOLUB (1997) 'Environmental policy and politics', in Rhodes, M. Heywood, P. and Wright, V. (eds) *Developments in West European Politics*, pp. 226–43, Basingstoke: Macmillan.

Comité 21 (1996) *Le Développement durable?: 21 entrées, soixante-quinze initiatives concrètes en France*, Paris: Comité français pour l'environnement et le développement durable.

Commission française du développement durable (1996) *Le Développement durable. Contribution au débat national*, Paris: CFDD.

B. DALAL-CLAYTON (1996) *Getting to Grips with Green Plans: Recent Experience in Industrial Countries*. With contributions from Izabella Koziell, Nick Robins and Barry Sadler, London: Earthscan.

A. DOWNS (1972) 'Up and down with ecology—the "issue attention cycle"', *Public Interest*, 28, 38–50.

D. DRON (1995) *Environnement et choix politiques*, Paris: Flammarion.

F. DUCHÊNE (1996) 'French motives for European integration', in Bideleux, R. and Taylor, R. (eds) *European Integration and Disintegration*, pp. 22–35, London: Routledge.

P. FRANÇOIS (1996) *Une Ecotaxe communautaire: quels effets environnementaux, économiques et institutionnels?*, Paris: Rapports du Sénat, no. 210.

J. GOLUB (1996) 'Why did they sign? Explaining EC environmental policy bargaining', Florence: European University Institute, EUI Working Paper RSC 96/52.

F. HATEM (1991) 'A propos du plan vert français', *Futuribles*, 152 (mars), 69–73.

Institut français de l'environnement (1994) *L'Environnement en France 1994–5*, Paris: Dunod.

P. LASCOUMES (1994) *L'Eco-pouvoir: Environnements et politiques*, Paris: La Découverte.

M. LIGOT (1991) 'La transposition des directives communautaires en droit interne', Paris: Imprimerie nationale, Assemblée Nationale, Délégation nationale pour les CE, rapport d'information no. 2292.

G. MAJONE (1996) 'A European regulatory state?', in Richardson, J. J. (ed.) *European Union: Power and Policy-making*, pp. 263–77, London: Routledge.

Y. MÉNY (1995) 'Politiques publiques en Europe: une nouvelle division du travail', in Mény, Y., Muller P. and Quermonne, J. (eds) *Politiques Publiques en Europe*, pp. 335–42, Paris, L'Harmattan.

P. MOREAU DEFARGES (1996) 'La France, province de l'Union Européenne?', *Politique Etrangère* (Spring), 37–48.

OECD (1997) *Environmental Performance Reviews: France*, Paris: OECD

R. ROMI (1990) *L'Administration de l'environnement*, Paris: Editions de l'Espace Européen.

S. ROYAL (1994) 'Discours', in Prieur, M. and Doumbé-Billé, S. (eds) *Droit de l'Environnement et développement durable*, pp. 15–18, Limoges: PULIM.

A. SBRAGIA (1996) 'Environmental policy: the "push-pull" of policy-making', in Wallace, J. and Wallace, W. (eds) *Policy-making in the European Union*, pp. 235–55, Oxford: Oxford University Press.

D. SIMON and A. RIGAUX (1991) 'Les contraintes de la transcription en droit français des directives communautaires: le secteur de l'environnement', *Revue juridique de l'environnement*, 3, 270–332.

J. THEYS and L. CHABASON (1991) 'Le Plan national pour l'environnement', *Futuribles*, 152 (mars), 45–73.

Part II

Clean Technology and Integrated Environmental Management

7 Energy and the Environment

Geoffrey P. Hammond

Summary

The interaction between energy use and the consequent environmental impact from pollutant emissions is examined principally from the perspective of the industrial world, but also in the context of the need to achieve sustainable development at a global level. It is argued that it is necessary to study modern energy systems using a rational, analytical, and interdisciplinary approach. Ideas emanating from the study of thermodynamics and life-cycle assessment suggest constraints on the energy policy options available to the United Kingdom and other industrial countries. Methods of analysis considered include the application of prescriptive techniques, such as energy and exergy analysis, in a general systems framework. They are employed to highlight some of the limitations of the so-called 'normative' techniques of economic analysis. Patterns of energy and electricity demand in the newly liberalized energy market within the UK are contrasted with the availability of indigenous fossil fuel resources and their medium-term rate of depletion. In order to illustrate the complexity of the energy and environmental issues a number of controversial topics are discussed, including the potential contribution to global warming of greenhouse gas emissions from fossil fuel burning, as well as the role of nuclear power and the 'renewable' energy sources. In addition, the desirability of significantly improving energy and resource use efficiency is also reviewed in the light of new UK Government's evolving energy, environmental and transport policies.

I INTRODUCTION

Energy sources of various kinds power human development, but they also put at risk the quality and longer-term viability of human life as a result of unwanted 'second order' or side effects. In order to analyse these complex interactions it is necessary to take an overtly rational approach, which treats these problems from a systems perspective. Many of the side effects of energy production and consumption are 'hot' issues in contemporary society. They give rise to potential environmental impacts on a local, regional and global scale. Examples include the depletion of North Sea oil and natural gas resources, the generation of smog from urban road transport, acid rain formation via pollutant emissions from (primarily) fossil fuel power stations, the difficulty of long-term safe storage of radioactive wastes from nuclear power plants, and the possibility of the enhanced greenhouse effect from combustion-generated pollutants. For a proper understanding of such issues it is necessary to employ a range of analytical techniques that have been developed under the umbrella of several different scientific, engineering and social science disciplines. To do otherwise than adopt an interdisciplinary, systems approach is to run the risk of missing some important element in the broad environmental canvas, or become prisoners to the latest 'green' fashion.

The purpose of the present contribution is to outline the range of concepts and analytical techniques that can provide important insights into the area of energy and the environment. A critical approach is taken to the use of these methods. It is argued that a portfolio of techniques is needed to devise a rational strategy. The difficulties inherent in this field are illustrated via the examination of several presently controversial issues, including global warming, the role of nuclear power, and that of energy conservation or improvements in the efficiency of resource utilization more widely. These are considered within the context of the need to achieve sustainable development.

II   ANALYSING ENERGY USE AND ITS ENVIRONMENTAL CONSEQUENCES

Systems Thinking and the Whole Life Cycle

Energy systems pervade industrial societies and weave a complex web of interactions that affect the daily lives of their citizens. In order to properly analyse the energy and environmental consequences of changes in supply and demand of energy-intensive goods and services, it is necessary to take a holistic approach. This implies drawing the system boundary quite widely; the nation state for policy decisions within the competence of national governments (for example, those influencing urban transport and air quality), regional intergovernmental blocks like the European Union where co-operation is needed to tackle problems on a continental scale (such as acid rain deposition), and globally for problems of the potential magnitude of the enhanced greenhouse effect. It is also important to trace the whole life of products, their associated energy flows and pollutant emissions as they pass through the economy. A simplified model of energy flows in the United Kingdom is illustrated in Figure 7.1. Heat is wasted and energy is lost at each stage of energy conversion and distribution, particularly in the process of electricity generation. The schematic energy flow diagram hides many feedback loops in which primary energy sources (including fossil fuels, uranium ore, and hydro-electric sites) and secondary derivatives (such as combustion and nuclear-generated electricity) themselves provide upstream energy inputs into the 'energy transformation system'.

The latter is that part of the economy where a raw energy resource

Figure 7.1 *Simplified representation of the UK energy system*

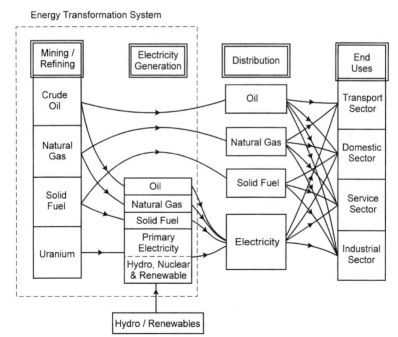

is converted to useful energy which can meet downstream 'final' or 'end-use' demand (Slesser, 1978). 'Renewable' energy sources are taken to mean those that are ultimately solar-derived: solar energy itself, biomass resources, and wind power. The energy resource input into the whole system is termed 'primary energy', and may be defined by:

Primary energy = downstream end-use or delivered energy
+ upstream waste heat

Consequently, the overall performance of an energy system can be represented in terms of an efficiency:

$$\text{Overall energy system efficiency} = \frac{\text{energy supplied to final consumers}}{\text{primary energy consumed}} \times 100\%$$

Despite significant changes in the fuel mix utilized within the UK energy system over the period from 1965 onwards, its overall performance has altered little (Hammond, 1998). The energy overheads (losses due to conversion inefficiencies and transmission losses) and overall system efficiency are given in Table 7.1. It can be seen that the efficiency for the supply to final consumers has varied by only some 2 per cent since 1965. This slight fall was primarily due to the greater use of electricity in the industrial sector, albeit supplied by new power plants with higher efficiency. Modern combined cycle gas turbine (CCGT) power stations have a thermal efficiency, for example, of 40–50% (Hammond, 1998).

Table 7.1 Overall performance of the UK energy system

	1965	1975	1985	1995
Primary energy consumed (PJ)	8,046	8,521	8,665	9,183
Energy supplied to final consumers (PJ)	5,662	5,893	5,940	6,318
Energy overhead (PJ)	2,384	2,628	2,725	2,865
Overall energy system efficiency (%)	70.4	69.2	68.6	78.8

Source: Hammond (1998); DTI, *UK Digest of Energy Statistics 1996*, tables A2.

The Limits to Economic Analysis

In the classic study of economics, 'micro-economics' as it is now known, the system studied is the individual firm or the consumer. This is a system that may well be sub-optimal in natural resource and environmental terms (Chapman, 1976). The transactions between the firm (or consumer) and the rest of the world are economically described in terms of the quantities and prices of the commodities exchanged. Prices in this classic economic model are supposed to reflect the 'value' that society places on an economic good. Thus, economics is claimed to be a 'normative' discipline: it suggests the optimal course action to be taken in the allocation of resources. For the price of commodities to give information that will lead to an efficient use of resources, it is necessary to assume that the prices are determined in a 'perfect competitive market'. Obviously there may be imperfections in the structure of the market; for example social costs may be excluded from prices, and 'externalities' (such as pollution or waste disposal costs) might not be included. There are also likely to be uncertainties about the future, restricted information about technological possibilities and time-lags, all of which might cause prices to deviate from those which would lead to optimal investment decisions.

The idea of prices reflecting economic value has been extended to the broader sphere of public sector or 'welfare' economics. This has led to the development of the techniques of cost-benefit analysis (Dorfman and Dorfman, 1993) for the assessment of public works projects. It now provides an important input into the evaluation of many projects that have significant impacts on the environment. These techniques play a major role in assessing new projects in the transport sector (see, for example, Maddison *et al.*, 1996), including road schemes and airport runways. In such cases it is necessary to internalize some of the costs and benefits that might otherwise be viewed as being external to the market. This valuation process is uncertain and potentially controversial, often relying on the determination of shadow prices. The costs and benefits in monetary terms are then progressively discounted for future years in order to allow for the 'time value of money'. Ultimately, the application of cost–benefit analysis (CBA) results in the determination of a single decision criteria (or a probability distribution for this criteria, if uncertainty is explicitly taken into account); typically the discounted cost–benefit ratio, or some related parameter.

There is no reason to think that the discount rate at which firms and consumers actually discount the future represents that rate at which society as a whole should discount it. Society may put a higher value on future resources than individuals and firms do. The latter will tend to consume natural resources faster than is warranted by society as a whole. Indeed, it may be argued that in the case of a depleting natural resource (such as fossil fuels or uranium) that perhaps a zero or even negative discount rate is appropriate (Munby, 1976). However, these have not, in practice, been used. It is certainly clear that discounted CBA techniques do not adequately reflect the resource depletion problem, at least as long as resource prices reflect mainly short-term trends. Nonetheless, the evaluation of social and environmental costs is obviously useful in identifying where the market has failed to internalize them. This provides governments with an indication of those areas in which action needs to be taken by way of the introduction of economic instruments (such as 'green' taxes) than can offset market deficiencies.

Some of the more ardent advocates of CBA techniques for evaluating new projects with significant environmental impacts imply that they can be used as the sole method of assessment. There are a number of reasons for discouraging such an approach. Firstly, the various methods for valuing external costs and benefits are all open to criticism. Choice of different valuation methods can lead to a wide variation in the supposed costs and benefits. In the extreme, they result in methods for valuing human life and well-being that are quite at odds with that perceived by individual people or society as a whole. Similar difficulties arise in valuing other elements of the biosphere. The second, and arguably more important, reason for discouraging the sole use of CBA techniques is that they obscure rather than highlight the range of impacts that may emanate from a given project. Decision makers are presented with a single decision criteria (such as the discounted cost–benefit ratio), which actually hides many disparate environmental impacts. It is vitally important that the implications of these impacts are faced, particularly by politicians, rather than obscured by the methodology.

Life-cycle Assessment

Life-cycle assessment (LCA) is the process of quantifying energy and materials used, and pollutants or wastes released into the environment as a result of a product or activity (Graedel and Allenby, 1995). It involves the entire life-cycle of the product or activity from 'cradleto-grave'. Unlike economic analysis, LCA is 'prescriptive' not 'normative'; that is it will indicate the consequences of actions rather than suggest optimal courses for action. The cost inventory may be employed, in part, as an 'input vector' to other analyses, such as energy analysis. This latter technique (sometimes termed energy or fossil fuel accounting) was developed in the 1970s in the aftermath of the oil crisis (see, for example, Chapman, 1976; Roberts, 1978 and Slesser, 1978). It was widely used by the academics and UK government departments, including the Energy Technology Support Unit (ETSU) at Harwell. However, it needs to be employed with some care as neither the whole life or 'gross' energy requirement (GER) may necessarily be the most appropriate criteria for assessing energy-related projects. It takes no account of the energy source in a thermodynamic sense. Electricity may be regarded (see below) as a high-grade source having a higher quality, or exergy, because it can undertake work. In contrast, low temperature hot water, although also an energy source, can only be used for heating purposes. This distinction is very important when considering a switch, for example, from traditional internal combustion (IC) engines to electric or hybrid vehicles. Thus, it is important to employ exergy analysis (see, for example, Szargut *et al.*, 1988) alongside a traditional energy analysis in order to illuminate these issues. Finally, the environmental life-cycle sequences should be examined, particularly in terms of various pollutant emissions: an approach known as 'environmental impact assessment' or EIA (Canter, 1996).

Energy analysis

In order to determine the primary energy inputs needed to produce a given output of product or service, it is necessary to trace the flow of energy through the relevant industrial system. This is based on the First Law of Thermodynamics; the principle of conservation of energy, or the notion of an energy balance applied to the system. The system boundary should strictly encompass the energy resource in the ground (for example, oil in the well or coal at the mine), although this is often taken as the national boundary in practice. Thus the sum of all the outputs from this system multiplied by their individual energy requirements must therefore be equal to the sum of inputs multiplied by their individual energy requirement. The process consequently implies the identification of feedback loops such as the indirect, or 'embedded', energy requirements for materials and capital outputs.

Figure 7.2 *Schematic representation of the energy analysis procedure (adapted from Slesser, 1978)*

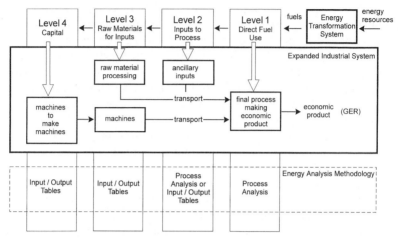

This procedure is illustrated schematically by Figure 7.2, which has been adapted from one given by Slesser (1978). Different 'levels of regression' may be employed depending on the extent to which feedback loops are accounted for, or the degree of accuracy wanted. The procedure leads to an estimate of the GER, sometimes loosely termed the primary energy 'cost'. It can be used to determine the least energy-intensive industrial process from amongst a number of alternative options.

There are several different methods of energy analysis; the principal ones being statistical analysis, input/output table analysis, and process analysis. The first method is limited by available statistical data for the whole economy or a particular industry, as well as the level of its disaggregation. Statistical analysis often provides a reasonable estimate of the primary energy cost of products classified by industry. However, it cannot account for indirect energy requirements, or distinguish between the different outputs from the same industry (Roberts, 1978). The technique of input/output table analysis, originally developed by economists, can be utilized to determine indirect energy inputs and thereby provide a much better estimate of GER. It is only limited by the level of disaggregation in national input/output tables. Process energy analysis is the most detailed of the methods, and is usually applied to a particular process or industry. It requires process flow charting using conventions originally adopted by the International Federation of Institutes of Advanced Studies in 1974–1975 (Chapman, 1976; Roberts, 1978 and Slesser, 1978). The application domain of these various methods overlap as is illustrated in Figure 7.2.

The techniques of 'first law' energy analysis have been widely used since the first oil crisis of the early 1970s. It was recently used by da Costa *et al.* (1997) to determine the life-cycle primary energy consumption of a modern, mid-range passenger car in the UK. They estimated the energy consumed when the vehicle was powered by alternative prime movers, specifically electric and parallel hybrid cars, compared to a 'baseline' petrol or diesel engine (see Figure 7.3). Although the electric vehicle appears the least energy-intensive option, versatility requirements for owners needing both urban and inter-city travel might dictate support for hybrid vehicles. Similar results were recently obtained by researchers at ETSU for light goods vehicles (Gover *et al.*, 1996).

Figure 7.3   *Life-cycle energy analysis of a passenger car with alternative prime movers (da Costa et al., 1997)*

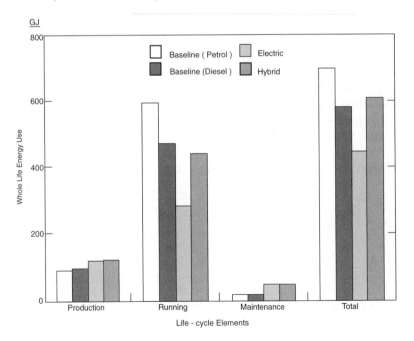

Exergy analysis

First law energy analysis as described above enables energy or heat losses to be estimated, but gives no information about the optimal conversion of energy. In contrast, the Second Law of Thermodynamics shows that not all the energy input into a system can be converted into useful work. It therefore provides the basis for the definition of parameters that facilitate the assessment of the maximum amount of work achievable in a given system with different energy sources (Wall, 1987, 1990; Szargut *et al.*, 1988; van Gool, 1997 and Winterbone, 1997). Exergy is the available energy for conversion from a donating source with reference to a specified datum; usually the ambient environmental conditions (typically 5–25°C). In a sense it represents the thermodynamic 'quality' of an energy source, and that of the waste heat or energy lost in the reject stream. Second law or exergy analysis is the only way of determining where the thermodynamic value of the source is lost in processes or in society. It provides a basis for defining an exergy efficiency, and can identify the 'improvement potential' within systems.

   The use of the exergy method of analysis has grown rapidly since the mid-1970s, particularly for optimising individual energy conversion systems and process plant (Szargut *et al.*, 1988; Winterbone, 1997). Van Gool (1997) cites the example of a gas boiler plant for space heating, which is normally regarded as having quite a high energy efficiency, but is found to have a very low 'second law' or exergy efficiency. More recently it has been applied to examine the overall performance of national energy systems. Van Gool (1997) illustrated the technique by reference to a hypothetical country ('Otherlands'), while Wall (1987, 1990) applied it successively to Sweden and Japan.

Whole life environmental impact assessment

The methodology of LCA in terms of the impact of environmental pollutants has only relatively recently been codified. This was done under the auspices of the Society of Environmental Toxicology and Chemistry (SETAC) at a series of workshops in the early 1990s (Graedel and Allenby, 1995). In the USA it is sometimes referred to as 'resource and environmental profile analysis', or REPA (Canter, 1996). The methodology follows closely that developed for energy analysis (Slesser, 1978), and evaluates the environmental burdens associated with a product or process over their whole life-cycle. This requires the determination of a balance or budget for raw materials (inputs), wastes and other environmental releases (outputs) emanating from the system. (Energy is often treated concurrently, although it was considered separately here.) The aim of the LCA is to identify opportunities for environmental improvements, by way of reducing the burdens. It follows a logical sequence of goal definition and scoping, inventory analysis, impact assessment, and recommendations for improvement. There are many technical issues that need to be addressed while conducting this type of LCA (and about which Ayres (1995) and Lee *et al.*, (1995) have been particularly critical), the definition of system boundaries, the quality of data available, and the way in which the results are normalized. The inventory analysis stage is quite well defined, although refinements to the methodology are frequently appearing in the literature. However, the impact assessment stage is far less certain, particularly in terms of the evaluation of impacts. Here it has much in common with CBA, except that it does not attempt to value the impacts in money terms or present them as a single decision criteria. Instead decision-makers are provided with, for example, a matrix of potential environmental impacts.

The general increase in environmental concern over the last decade has led naturally to the need for assessment techniques, such as LCA and EIA. These have been particularly widely adopted in the areas of chemical process engineering and land use planning. The study of the energy performance of light goods vehicles with alternative prime movers by ETSU (see above, Gover *et al.*, 1996) was accompanied with a LCA to determine the whole life vehicle emissions. Similarly, Nuclear Electric plc (1994) undertook a 'full fuel cycle' assessment of various power station types. The results for CO_2 emissions are shown in Figure 7.4 (see also Hammond, 1997a), where the benefits of nuclear power in this regard are obvious. They also produced comparable emissions inventories for SO_2 and NO_x. A conclusion that can be drawn from such studies is that moving to modern, cleaner technologies, whether they be motor vehicles or power stations, tends to reduce the emission of most pollutants. Thus, measures that are aimed at lessening SO_2 emissions (a major acid rain precursor) have a knock-on benefit of lowering CO_2 and other pollutants. However, this generalization should not be taken too far, and individual cases need to be properly assessed using the sort of techniques outlined here.

Figure 7.4　*Carbon dioxide emissions attributable to UK electrical power generation (adapted from Nuclear Electric plc, 1994)*

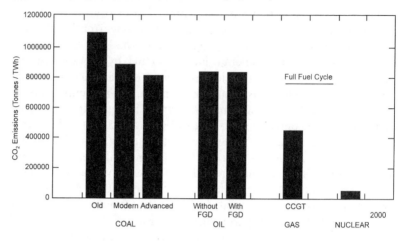

III　UK ENERGY BALANCES

Electricity as an Energy Source

Electricity is a high-grade energy source in the sense that it can be used to provide either power or heat. It is essentially a 'capital' resource that can be generated using either depleting fossil or nuclear fuels (Hammond, 1997b). These latter sources may be contrasted with the renewable (or 'income') energy sources, such as solar energy and tidal, wave and wind power. However, large energy losses occur in generation unless used in conjunction with combined heat and power (CHP) systems. It is also wasteful in thermodynamic terms to convert fuels to electricity only to employ them for heating. If heat is required, then it would be far more efficient to burn fossil fuels (for example) to produce heat directly. Chapman (1976) discusses the relative end-use merits of electricity, arguing that (in spite of the lack of detailed statistics) it was possible to estimate that some 25–35% of electricity in the UK was used for heating in the mid-1970s. Lovins (1977) has suggested that in the USA only some 10% of final energy use requires electricity. In any case, the latter is a rather poor substitute in transport, which predominately uses petroleum products. The issue of substitutability of fuels in the transport sector is still one that is insufficiently addressed given some of the projections for the short life of oil and natural gas (see below, and Hammond, 1998).

Another limitation of electricity generation, certainly in terms of meeting global energy demand, is its requirement for a high-technology infrastructure. It is often argued that energy demands will grow in the future to meet the rising needs and expectations of the developing countries, who often have rapidly growing populations. However, these countries are unlikely to have the financial resources or expertise to follow a high-technology route (Hammond, 1997b) certainly not in Africa (except South Africa) or in much of Latin America.

UK Indigenous Fuel Balances

In the immediate post-Second World War period, most fossil fuel reserves were plentiful in the 'West' (the member countries of the Organization of Economic Co-operation and Development, OECD), with cheap petroleum supplies readily available from the Middle East. Solid fuel (coal, coke and breeze) were the main source of energy for home heating, in industry, and for electricity generation. Oil dominated the road transport sector, where it could not easily be substituted by alternatives, but also began to encroach on the domain of 'King Coal'. It was coal that fired the economic recovery of Europe in the 1950s, which in turn led to increased imports of oil in order to match the demands of the 'economic miracle' (Andrews, 1991). The relatively high price of coal and moves to encourage the use of 'smokeless' fuel on environmental/health grounds, particularly in the home, induced coal/oil substitution in the domestic, industrial and services sectors.

The growth in global petroleum consumption throughout the 1960s contained the seeds of instability in the oil market, with two-thirds of supplies being imported from the Arab countries (Andrews, 1991 and Hammond, 1998). The major destablishing event occurred in October 1973, when Israel was attacked by its Arab neighbours in what became known as the Yom Kippur War. The Middle East members of the Organization of Petroleum Exporting Countries (OPEC) felt that the United States and some of its European allies had overly supported Israel. In retaliation Saudi Arabia imposed an oil embargo on the USA and the Netherlands, which ultimately led to a cut of 17% in OPEC's output. This induced the first oil price 'hike' or shock, with prices rising from US$3 per barrel to $11 over the period of a few months. The impact of these events was to engender a feeling of inse-curity in Western industrialized countries, and they collectively started to introduce measures that would free them from effective depen-dence on OPEC oil supplies. These were co-ordinated by a new body, the International Energy Agency (IEA), established under the umbrella of the OECD. On the supply-side, encouragement was given to the development of non-OPEC, particularly indigenous, sources of energy. This was coupled with a variety of energy efficiency measures. In the medium-term (5–10 years), these were spectacularly successful in ending the reign of OPEC as the oil price setter.

Britain played its full roll during this period, in consort with its part-ners in the IEA/OECD. In 1967 the UK commenced piping ashore natural gas from the West Sole field of the southern North Sea basin (Andrews, 1991). It marked the start of a major investment pro-gramme, which accelerated throughout the 1970s, aimed at convert-ing every gas burning home from coal (or 'town') gas to natural gas. This eventually induced a major shift in consumer preference towards gas-fired central heating. Natural gas consequently increased its share of the domestic energy market from some 13% in 1965 to just over 65% by 1995 (Hammond, 1998). Exploration for oil as well as natural gas progressively moved out of the shallow depths of the southern North Sea to the deep waters between Scotland and Norway. Here a number of giant oil fields were discovered over the period 1969–1972. Development of North Sea oil during the 1970s and 1980s led to the UK becoming self-sufficient in petroleum supplies by 1981, as shown in Figure 7.5 (updated from the original work of Hammond and Mackay, 1993). This comparison of indigenous oil supply and demand was based on a skewed normal production profile for the UK conti-nental shelf. The area beneath the projected supply curve represents the ultimate recoverable reserves, although Hammond and Mackay (1993) noted that these were heavily dependent on geological esti-mates which vary significantly over time. They employed data up to 1988, and Figure 7.5 therefore illustrates the short-term accuracy of their forecasts. In the intervening period oil demand has followed their projections quite closely, but North Sea production has out-stripped the forecast. Similar projections for UK natural gas supply and demand are shown in Figure 7.6. Here it can be seen that the so-called 'dash for gas' has outstripped most contemporary predictions. Even these recent estimates of Hammond and Mackay, based on EU forecasts published in 1990 (Commission of the European Commu-nities, 1990), under-predicted by some 85% the electrical output of

Figure 7.5 *UK indigenous oil balances (updated from Hammond and Mackay, 1993)*

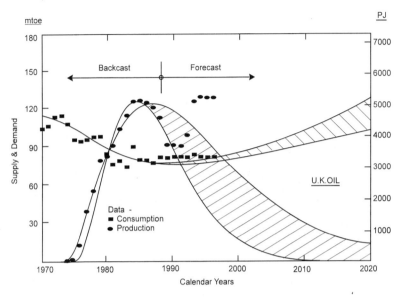

new CCGT plants in 1995. Thus, by early 1997 the dominant position of coal-fired power stations in UK electricity supply had changed dramatically from having held 60% of the market to only 33% in just 5 years. They had been overtaken by nuclear power plant as the largest generator with a 36% share, followed by natural gas which captured 29% of the market. The new Labour Government in Britain continues to give planning consent for new CCGT plant, with a preference for their incorporation into CHP schemes.

Energy Market Competition

The year 1979 saw the election of a Conservative Party Government in Britain led by Margaret Thatcher. It was ideologically committed to a process of 'privatization' of state-run industries which dominated the energy sector. These energy utilities had been regarded hitherto as 'natural monopolies', based on the argument that it was not practical to lay more than one gas pipeline or electrical cable network. The UK oil industry had always been a competitive market, while in contrast, coal, natural gas, and electricity supplies were all state

Figure 7.6   *UK indigenous natural gas balances (updated from Hammond and Mackay, 1993)*

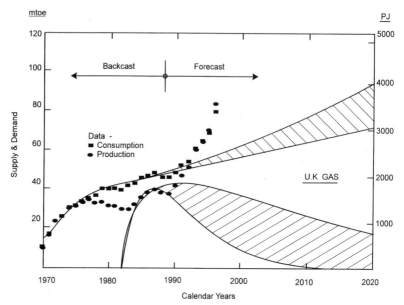

monopolies. The Conservative Government set about successively privatizing these utility companies, with a view to introducing private capital and reducing the public sector borrowing requirement. British Gas was the first to be floated on the London Stock Exchange in December 1986. It retained a monopoly position in respect to the non-industrial of 'tariff' market. The company and the market more broadly were overseen by a Government regulator, who had powers to limit the company's prices and profits. Here the argument seemed to be that the process of introducing private capital, together with competition against alternative energy sources, would induce efficient management to the ultimate benefit of the consumer (Hammond, 1998). This notion did not last long; the electricity industry was subsequently broken up into 12 distributors and 3 generators (in England and Wales; Scotland and Northern Ireland having a rather different structure), while competition in the tariff gas market will be introduced during 1996–1998.

The electricity utilities were floated over the period 1990–1995, with the exception of the nuclear power generators. Privatisation proved to be an uncomfortable experience for the UK civil nuclear electricity industry. It was found that Nuclear Electric and Scottish Nuclear could not be floated with the other generators in 1991. The Stock Exchange had grave concerns over the costs of nuclear-generated electricity, and in particular the costs associated with decommissioning their older plants. When the Government eventually floated part of the nuclear industry in 1996 (as British Energy), the liabilities of the old Magnox reactors remained in state ownership. One of the first actions of the privatized company was to announce that it would cease for the time being plans to construct further nuclear generating capacity, including the Sizewell C scheme. They did not, however, exclude the possibility of investing in other types of plant, such as CCGT generators. This was a reflection of the relative costs of generating electricity from nuclear and CCGT plants (Hammond, 1997b).

Liberalization of the energy market was completed by the competitive sale of the British coal mining industry in 1994. About 70% of the output (80% of deep-mine production) was sold to one company; RJB Mining. At that time major, medium-term contracts were in place to supply the non-nuclear electricity generators, although these will only be renewed on a competition basis. Imports of cheap coal from abroad are now permitted, and by 1995 these amounted to some 20% of indigenous production. The spot price for coal fell during the period 1981–1994 from US$75 per tonne to $40 in current (money-of-the-day) terms. The relatively cheap price of natural gas from the North Sea, and the low capital cost of CCGT plant, meant that this sector of the electricity supply industry expanded rapidly.

On the demand side, the energy policy of the former Conservative Government in Britain was aimed at letting competitive energy prices act as the main driver for energy efficiency measures. In other respects it concentrated on the provision of energy saving advice to the domestic, commercial and institutional (or services), and industrial sectors. An Energy Saving Trust was also established with support from British Gas and the electricity utilities in 1994 to promote efficient use of energy by domestic and small business consumers. The effect of these policies, and of structural changes in the economy generally over the period 1979–1997, can be seen in the breakdown of UK energy demand by final users illustrated in Figure 7.7. This is taken from a retrospective study by Hammond (1997a, 1998) of energy projections made in the later 1970s by various forecasters. The projections shown in Figure 7.7 are those of Gerald Leach *et al.* (1979) at the International Institute for Environment and Development (IIED) in

Figure 7.7 *UK energy consumption by final user sectors (Hammond, 1998)*

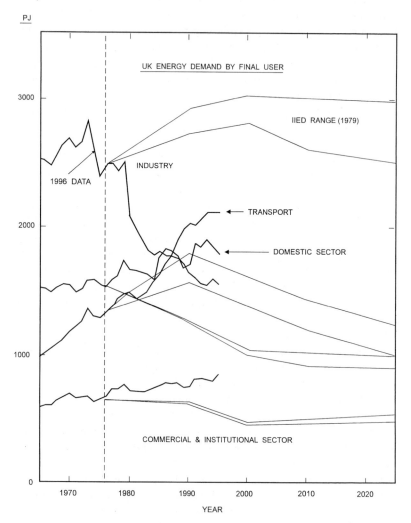

London. Domestic sector delivered consumption is seen to have remained relatively stable at a share of about 26–30% since 1965, while the commercial and institutional sector demand has risen modestly from just over 10% to nearly 14%. It is within industry and the transport sector that major changes have taken place. Industrial energy demand has fallen from a share of 45% to just below 25% over this 30-year timeframe. This is largely due (Hammond, 1998) to structural changes, with a move away from heavy industries, and energy efficiency gains induced by the energy price mechanism. Transport's share of energy consumption has grown from some 18% to 34%, matching the increase in passenger and freight road traffic.

IV   ENERGY AND SUSTAINABILITY

Sustainable Development

In the aftermath of the 1992 Earth Summit in Rio de Janeiro (the UN Conference on Environment and Development) there has been a greater awareness of the need to devise strategies aimed at sustainable development (Hammond, 1997b); balancing economic and social development with environmental protection. The World Commission on Environment and Development (1987) in the influential Brundtland Report defined sustainable development as meeting 'the needs of the present without compromising the ability of future generations to meet their own needs'. Hence, governments have been encouraged to conserve depleting fuel resources, and to make greater use of renewable energy sources. However, the World Energy Council (1993) has suggested that renewables can only contribute some 4–5% of total energy consumed to 2020. This has led to a debate between the advocates of renewables and those favouring nuclear power. Both have argued that the concept of sustainability supports their widespread adoption. Thus, John Collier (no date), the late Deputy Chairman of British Energy, suggested that nuclear power is one means of helping to maintain the energy resource balance of the planet at the same time as safeguarding its environmental quality. He had the issue of greenhouse gas emissions and global warming specifically in mind (see below). On the other hand, Greenpeace (1994) argue that nuclear power will leave future generations with a legacy of nuclear waste. The two positions could, of course, be reconciled if long-term safe storage of waste could be assured, but this is difficult to prove *a priori* (Hammond, 1997a).

The combustion of finite reserves of fossil fuels results in their obvious depletion over time. It is often argued, particularly by economists, that the carrying capacity of the planet as a whole is so large that new discoveries offset current production. Although this may be true on a global scale, it is unlikely to apply at the level of the individual nation-state, or even at a regional scale (such as within the European Union). There is considerable uncertainty over fossil fuel resources in the medium term. The lifetime and global distribution of these vary enormously (Hammond, 1997b):

* Oil: OPEC (Middle East)-dominated, 20–40 year life
* Natural Gas: CIS (Russian)-dominated, 40–70 year life
* Coal: Widely distributed, 80–240 year life

These figures are rough estimates assuming current rates of consumption (BP, 1997), but they indicate that the sources of fossil fuel supplies for OECD countries, with the exception of coal, are rather insecure. If depletion of oil and gas at anything like this rate actually occurred, then the price of these fuels would rise. This would make the financial case for renewable energy sources and nuclear power look much brighter. It has often been argued since the 'oil crises' of the 1970s that a nuclear power and/or renewable energy strategy should be adopted as an 'insurance policy' against the insecurity of the oil market. In reality these resources are not substitutable, particularly in the transport sector (Hammond, 1997b).

In order to ensure sustainability of energy supplies, particularly in regard to the needs of future generations, it is necessary to conserve non-renewable resources and exploit the potential of renewables. The current fuel mix in the UK is heavily dependent on fossil fuels, which are being depleted at a fast rate; see Tables 2 and 3 taken from Hammond (1998). The ratios of reserves to production of coal, oil and natural gas in Britain are all much lower than the global figures indicated above, and exhibit an alarming downward trend. There is still considerable uncertainty over the magnitude of indigenous reserves of oil and gas. Nevertheless, even with new discoveries (such as the West of Shetland or 'Atlantic Frontier' oil province), it is likely that the UK will soon return to the position of being a net importer with deleterious effects on the balance of payments, and economic growth generally. Certainly insufficient attention is being paid by British politicians to the consequences of the depletion of North Sea resources.

The switch from coal to natural gas as a result of the introduction of a competitive energy market in Britain has had a favourable short-term effect on greenhouse gas emissions. Consequently, the Government will meet the internationally agreed carbon dioxide (CO_2)

Table 7.2   UK supply and demand of primary energy resources (PJ)

	1965	1975	1985	1995
Indigenous production	5,404	5,171	10,572	11,299
Imports	3,818	4,847	3,128	3,274
Exports (shipments)	–817	–774	–4,565	–4,955
Stock changes: balancing term	–71	–294	–84	+199
Non-energy use ('petro-chemicals')	–299	–429	–386	–634
Inland energy consumption	8,046	8,521	8,665	9,183
UK net production/ consumption	0.297	0.155	1.341	1.344

Source: Hammond (1998); DTI, Digest of UK Energy Statistics 1996, tables A11, A12.

Table 7.3   UK indigenous fossil fuel resources

Fuels	1965	1975	1985	1995
Solid fuels				
Reserves, R (10^6 tonnes)	–	–	10,750	2,500
Production, P (10^6 tonnes/year)	191.6	128.7	94.0	52.6
R/P ratio (years)	–	–	114.4	47.5
Petroleum				
Reserves, R (10^9 barrels)	–	16.0	13.0	4.3
Production, P (10^6 barrels/year)	0.6	12.1	984.7	1,005.6
R/P ratio (years)	–	1,322.3	13.2	4.3
Natural Gas				
Reserves, R ($10^{12}\,m^3$)	–	1.4	0.9	0.7
Production, P ($10^9\,m^3$/year)	0.2	35.4	39.7	71.5
R/P ratio (years)	–	39.5	22.7	9.8

Source: Hammond (1998); BP Statistical Review of World Energy, 1996; DTI, Digest of UK Energy Statistics 1996.

emission targets for the year 2000 without having to take any conscious action. This initial target was aimed at returning emissions to 1990 levels by the end of the decade. It became evident as early as 1993 that this would be readily achieved, although the UK is likely to agree to a further reduction of 15–20% for the year 2010 at the third session of the Conference of Parties to the UNFCCC in Kyoto, Japan

(December 1997). This will require significant governmental intervention, over and above falls in emissions stimulated by energy market liberalization. However, the 'free' market is unlikely to achieve a satisfactory balance between the broader requirements of sustainable development and environmental protection over the longer term. The short time horizon of markets means that they tend to ignore both the depletion of fossil fuels and the long-term potential for technological innovation; 'pinchpoints' often go unrecognized by commodity and stock markets until they are almost reached. In addition, as the British Government's own (advisory) Panel on Sustainable Development (1997) has recently argued, energy prices do not account for 'externalities', including the environmental costs associated with damage to the natural environment and possible climate change. One way of satisfying conflicting short- and longer-term needs might be to extend the regulatory framework for the UK energy utilities to include sustainability criteria. At the time of writing (Autumn 1997) the new Labour Government has instigated an inter-departmental review of utility regulation, including the electricity and natural gas markets. The focus of the review will be on the interests of consumers, although issues relating to sustainable development will also be considered.

Carbon Dioxide Emissions and Global Warming

Climate change is a complex phenomenon at a detailed level, and it is still poorly understood even by scientific 'experts' (Houghton *et al.*, 1996, 1997). However, the basic principles behind the Earth's heat balance are quite simple. The planet is covered by a thin atmospheric blanket that protects it from the severity of incoming solar, or short-wave, radiation. Trace gases in the upper atmosphere, and clouds nearer the surface, reflect a proportion of the Sun's rays back out into space. That which gets through is absorbed by the land or the oceans; thereby heating them up. These warmer surfaces in turn re-radiate infrared, or long wave, thermal radiation back to the atmosphere. Some of the trace gases then absorb, or trap, this heat in much the same way as glass does in a greenhouse. Consequently, these radiation absorbing gases are known as 'greenhouse gases'; the most important of which is carbon dioxide. This energy balance gives rise to the surface air temperatures that humans experience. It is hotter at the equator because the region generally faces the Sun directly, and vice versa for the poles. Seasons occur due to the tilt and rotation of the Earth about the Sun.

Energy use by humans to power our industrial society gives rise to the emission of both waste heat and extra carbon dioxide. This results from the burning of fossil fuels, such as coal, oil and natural gas, in power stations, motor vehicles, and heaters of all kinds. The fear is that anthropogenic (human-induced) greenhouse gases are building up in the atmosphere, and will lead to enhanced warming and large-scale climate change. It could, so the pessimists argue, lead to a range of damaging climatic upheavals. Glaciers may melt which, together with the expansion of warmer sea water, can give rise to the flooding of island and coastal communities. It is thought that the wider claims of possible melting of the polar ice caps will not happen (at least during the early stages of greenhouse gas warming), due to greater snow fall at the poles. In contrast, agricultural regions across the world may become drier and less productive, with droughts in already deprived regions. Such changes are likely to be geographically varied, and other regions might well become wetter and more windy. Indeed, the main consequence of global warming may initially be an increase in the occurrence and severity of extreme climatic events.

It is the details of the interactions between different elements in the climate system that scientists have difficulty understanding (Houghton *et al.*, 1997). They employ sophisticated computer models of the atmosphere but, at the present state-of-the-art, they are unable to accurately predict the impact of the enhanced greenhouse effect (Carson, 1997). One of the leading groups in climate change research is the Hadley Centre at the Meteorological Office in Bracknell. It advises the UN's Intergovernmental Panel on Climate Change (IPCC), as well as the UK's Department of the Environment. They employ global circulation models originally developed for long-term weather forecasting, which are run on some of the world's fastest and largest computers. In order to apply such models to examine human-induced climate change, many new interactions have to be incorporated. For example, it was only in 1995 that the Hadley Centre simulated the influence of sulphate aerosols, resulting from the emission of sulphur dioxide (again from fossil fuel combustion). They found (Hadley Centre, 1995) that these aerosol particles had a cooling effect; actually reducing predicted surface air temperature changes by about one third compared to their earlier forecasts with carbon dioxide alone (see Figure 7.8). This is a large difference. The recent

Figure 7.8   *Hadley Centre global climate model predictions (adapted from Hadley Centre, 1995)*

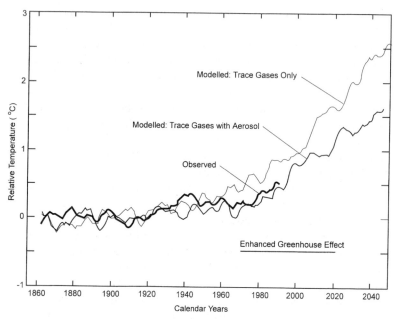

computations also indicated significant regional variations around the Earth; highlighting the fragility of the models. But sulphate aerosols are not the only phenomena that needs to be introduced into climate change simulations. Other important elements of the carbon cycle (exchanges between the atmosphere, oceans and biosphere) and the full effect of atmospheric chemistry will not be incorporated into the Hadley models until nearer the turn of the century (Carson, 1997). The Centre's research scientists are suitable circumspect in regard to the uncertainties inherent in their work, although these misgivings don't seem to reach the wider public domain.

   The current version of the Hadley Centre's climate model can broadly track the upward trend in global surface air temperatures over the last one hundred years or so (see again Figure 7.8). However, the Earth's climate is highly variable, and it will be some years before greenhouse gas warming emerges from the background variability. Localized events, including the generally drier weather in the UK, or the thinning of the Antarctic polar ice sheet, are not in themselves conclusive proof of an enhanced greenhouse effect. In any case, there may be some positive benefits from anthropogenic carbon dioxide emissions, which act as a fertilizer and may actually stimulate agricultural production in some areas. It is being argued by the IPCC and others that a consensus of scientific opinion supports the existence of the enhanced greenhouse effect, and believe that its impact will be very damaging in the next century. This assertion is in itself worrying, as it is incumbent upon the research community to exhibit a critical approach to fashionable theories. Scientific truth and certainty are not natural bedfellows at the frontiers of knowledge.

An attractive hypothesis for climate change that is gaining some scientific support is the possibility that the recently observed upward trend in surface temperature may be linked to variations in the Sun's output. Small, but climatically significant, changes in the 'solar constant' appear to be connected with sunspot activity on a 10–12 year cycle. Groups working in both Europe (Eigil Friis-Christensen and his co-workers at the Danish Meteorological Institute) and North America (principally Judith Lean at the Naval Research Laboratory in Washington and her collaborators) have suggested that this coupling could be more significant than the enhanced greenhouse effect. An early representation of the relation between the sunspot cycle and relative temperature anomalies (deviations from the 1951–1980 mean) in the Northern Hemisphere land air temperature (Friis-Christensen and Lassen, 1991) is shown in Figure 7.9. These datasets apply to the period 1891–1989, although Lean (1997) gives similar results for solar irradiance extending back to nearly 1600. Criticisms of the solar forcing connection have centred on (i) the absence of a plausible mechanism linking it to temperature anomalies, and (ii) the apparently weak statistical correlation. However, Svensmark and Friis-Christensen (1997) recently argued, based on satellite observations, that variations in global cloud cover (3–4% during the latest solar cycle) are strongly correlated with cosmic ray flux. This in turn is inversely proportional to sunspot activity. It is known that the distribution of cosmic rays emanating from the solar wind is modulated by the Earth's magnetic field. These rays may, according to Svensmark and Friis Christensen, generate electric charges in the atmosphere which could interact with sulphate aerosols and thereby stimulate cloud formation. Thus, the physical connection could be made between the solar cycle, variations in cloud cover, and the consequent alteration to surface temperatures. These results indicate that solar forcing can account for some 75% of global warming after the Little

Figure 7.9   *The solar link to global warming (adapted from Friis-Christensen and Lassen, 1991)*

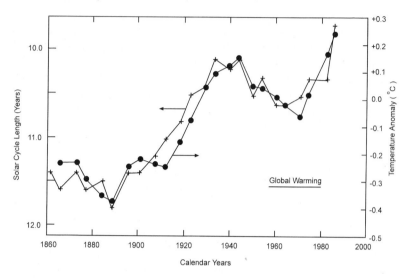

Ice Age of the late 1600s. Nevertheless, estimates of the influence since 1970 vary from as little as 10% too as much as 50% of the observed warming (Kelly and Wigley, 1992; Kerr, 1996; Calder, 1997), with the enhanced greenhouse effect contributing the remainder. This alternative model of climate change has not found favour among the leading scientific advocates of the anthropogenic global warming associated with the IPCC; although it has recently been popularized by the science journalist Nigel Calder (1997). It certainly deserves serious consideration, given the uncertainties in the enhanced greenhouse effect outlined above. At the very least, it may 'buy time' for humanity to devise a rational strategy for achieving sustainable development.

Nuclear Power and the Environment

One motivation for the construction of civil nuclear power plants from the early 1960s was that they could utilise fissile material produces by the military nuclear weapons programme, initially in the USA, USSR, Uk and France. They were consequently foreseen by governments as being part of a wider strategic development. By the end of the 1960s, civil nuclear power programmes had evolved a life of their own. The peaceful use of nuclear electricity generation has now spread to such an extent that some 30 countries operate nuclear power stations. However, by the 1980s people in the West became far more aware of the possible disadvantages of nuclear power encapsulated in Amory Lovins' phrase 'the Hard Energy Path' (Lovins, 1977). Two events reinforced this perception: the reactor failure at Three Mile Island (USA) in 1979 which was contained, and that at Chernobyl (USSR/ Ukraine) in 1986 which was not. In addition, many people became concerned about handling radioactive materials, principally radioactive emissions during the fuel cycle, the subsequent disposal of high- and intermediate-level wastes, and the eventual need to decommission plants. The storage of high-level waste and decommissioning of plants would require facilities which would operate safely over many decades. Another factor that discouraged nuclear power development is that the prospect of cheap nuclear energy seemed much less certain than had been claimed by its early advocates. The capital costs are high and they have not been offset by low running costs, particularly if waste storage and decommissioning costs are taken into account. The private sector has generally proved unwilling to meet these liabilities without government financial support in one form or another.

The gloomy aspects for nuclear electricity generation in the 1980s have been partially transformed in the 1990s. This has come about owing to the realization that there may be serious global environmental consequences of the continued burning of fossil fuels. Combustion of these fuels produces carbon dioxide, which many believe will contribute to global warming, leading to possible serious climate change (but see the discussion above). The amount of CO_2 produced by electricity-generating plant using different fuels is illustrated in Figure 7.4, where the benefit of nuclear power is clear. Similarly, the fossil fuel exhaust gases that contribute to acid rain (sulphuric dioxide and the nitric oxides) are also virtually eliminated by the nuclear fuel cycle (again, see above). In addition to the problem of pollutant emissions from fossil fuel combustion, there is also now a better awareness of the finite nature of oil and natural gas reserves. The OECD governments are especially anxious about the global distribution of these energy resources, which in the next 20–40 years will be located largely outside their control. If depletion of oil and gas at anything like the rate indicated above actually occurred, then the price of these fuels would rise. This would make the financial case for nuclear energy look much brighter. It has often been argued since the oil crisis of the

1970s that nuclear power should be adopted as an insurance policy against the insecurity of the oil market. In reality, the two resources are not substitutable, particularly in the transport sector (as discussed above).

Perhaps the main factor that will determine the extend to which Western Europe will embrace the nuclear option to meet a significant fraction of its electricity needs into the next century is its public acceptability. This is determined by the degree to which the public are convinced that nuclear power stations can operate safely, and that radioactive by-products can be securely stored over long periods. The extent to which people in different European countries appear to accept the nuclear option varies quite widely; presumable due in part to cultural factors. In France, which has very little in the way of indigenous fossil fuel reserves, the nuclear share of electricity generation is already about 73% (Hammond, 1997b). Certainly, the French have put up little resistance to this in contrast with other countries where there are active and vocal opponents of nuclear power, who have a significant influence on the public via their access to the media. The safety record of the nuclear energy industry compares favourable with its competitors (Fremlin, 1987), but is often perceived as being more life threatening; arguably out of proportion to the actual risks. Obviously, it is incumbent on the civil nuclear industry to reassure the general public over its long-term operating safety; a task that is undoubtedly daunting.

Trends in the generation of nuclear electricity from fission reactors have recently been critically analysed by Hammond (1997b) in terms of the main geopolitical of regional groupings that make up the world in the mid-1990s. This has been done by examining several recent, but somewhat conflicting, forecasts of the role of nuclear power in the fuel mix to about 2020 (see, for example, Figure 7.10). These indicate a growth in nuclear electricity generation world-wide of between 30 and 120% over the period 1990–2020. The major expansion in generating capacity is likely to take place in the Asia–Pacific Rim: the People's Republic of China, Japan, South Korea and Taiwan. These so-called 'Tigers' have rapidly growing economies with carrying capacity for a major expansion in nuclear power facilities. They typically have very limited reserves of indigenous fossil fuel supplies (except China). In addition, they are generally attracted by a high-technology energy option, and arguably have rather lower concern for enviromental protection than do European and North American OECD nations. Similarly it has been argued in some industrialised countries that it is

Figure 7.10   *World-wide nuclear electricity generation: data and projections (Hammond, 1997b)*

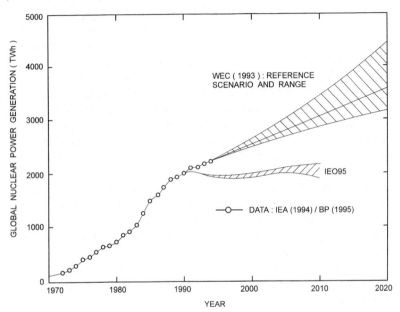

important to keep a technical capability in civil nuclear power, so that the West does not fall behind in this area of high technology. It is also seen as a technology with considerable export potential. Considerations of this type have certainly influenced the industrial strategy of countries such as France (and Japan). Thus the global nuclear industry overall will continue to be dominated by a small number of large nuclear electricity generating countries; principally the USA, France and Japan. These three industrialized states accounted for nearly 60% of installed nuclear capacity in 1990, and they are still likely to retain roughly this share by 2010 unless some of the more exaggerated projections are fulfilled (Hammond, 1997b).

Transport Energy Use and the Environment

It has become widely recognized that two of the main 'drivers' for change in the transport sector are energy use and its consequent environmental impact. These issues were highlighted in the UK, for example, by the Royal Commission on Environment Pollution

(RCEP, 1994, 1997) and the Technology Foresight Panels on Energy, the Environment, and Transport (1995a,b,c). They all emphasized the need to devise a sustainable transport system, although sustainability is difficult to achieve with the present mix of transport modes. This is because conventional vehicles are major consumers of non-renewable energy and material resources, while emitting pollutants that may do long-term harm to the environment. The transport sector relies on an energy network (petrol, natural gas, or even the electricity distribution system) that is national or even international. It currently accounts for some 35% of the consumption of petroleum products in the UK and very significant qualities of pollutants: carbon monoxide 90%, nitrogen oxide 60%, particulates 50%, and volatile organic compounds (VOCs) 40% (RCEP, 1994). Urban transportation systems obviously lead to traffic congestion and poor local air quality. Transport energy use in the UK has grown much more rapidly than in other demand sectors (see Figure 7.7) over the last 30 years. It also outstripped the increase in real GDP (Hammond, 1998). Indeed, the energy forecasts by Leach *et al.* (1979) were far too optimistic in their assumptions about the potential for saturation in car and air travel after 1990, and the rate of fall in vehicle specific fuel consumption.

The Automobile Association (1997) have argued that improvements in engine technology and fuel quality, together with the introduction of catalytic converters in the late 1980s, has resulted in a dramatic fall in 'tailpipe' exhaust fumes. Certainly the historic link between emissions such as NO_x, VOCs and particulates (PM10s) and road transport growth has been broken. This will undoubtedly lead to improved air quality and public health in city centres. However, energy use and CO_2 emissions have been largely unaffected by these developments (RCEP, 1994, 1997). Similarly, the continued rise in private car use, particularly in urban areas, will still lead to increased traffic congestion and noise. Consequently, the need for more radical technological and socio-technical changes are evident.

A large reduction in energy use and associated pollutant emissions could obviously be achieved via a major switch to public transport. However, the increased demand for personal mobility may still inhibit this in the foreseeable future; notwithstanding government and European Union consideration of various fiscal and land-use planning measures. The RCEP (1994, 1977) have advocated the development of cleaner vehicles and a move towards better integration with public transport systems. Travel by bus or train is much less energy-intensive than private car journeys, and consequently gives rise to far lower pollutant emissions per passenger-km. In order to encourage travellers to switch between modes, good public transport services must be provided along with easy interchange between modes. The Transport Technology Foresight Panel (1995a) suggested two targeted projects for the UK in this field; denoted as 'the informed traveller' and 'clear zones'. In the case of the first thematic area, the aim is to provide enhanced travel information services (using the latest developments in information technology) that will stimulate the use of public transport in the newly privatised bus and rail network. The clear zones initiative represents a coherent vision around which a variety of technological innovations could contribute to the evolution of more liveable city centres. It would focus on efficient private/public transport interfaces, and the use of control engineering techniques to optimise traffic flow and minimise pollutant emissions in urban areas. Proposals for improving the energy efficiency of road vehicles have tended to concentrate either on incremental technical fixes, or on ideas for more radically different design concepts. Environmental groups, on the other hand, have advocated more radical changes in lifestyle. Some groups in the UK (including SUSTRANS and 'Reclaim the Streets!') are committed to virtually eliminating the private car from inner cities in favour of walking, cycling, and cheap public transport.

Pressure on the multinational car companies to develop low emission motor vehicles has come from a variety of sources. In California the State Government has introduced measures to encourage these companies to introduce zero emission vehicles (ZEVs) into the market. A target of 10% of all vehicles by 2003 was originally set, although this has now been relaxed. In reality ZEVs imply electric vehicles with zero tailpipe emissions, but increased emissions at the power station. The Californian programme has driven research programmes into low emission vehicles, both by multinationals and the US Department of Energy. This has had a knock-on effect in Europe, although France has a considerable incentive to introduce electric vehicles because of the large fraction of electricity that is produced by base-load nuclear generation plant (currently a little over 70%; see above). Developments to date in the UK have tended to follow those elsewhere. These have been mainly relatively small-scale trials of alternative vehicles; principally small electric cars and vans, as well as large buses operating on compressed natural gas.

The agenda for radical design concepts in road vehicles has

arguably been set by Amory Lovins *et al.* with his vision of a 'hyper-car' (1993). This vehicle, termed 'supercar' in the original report, was postulated as being able to leapfrog incremental design improvements by adopting next generation technologies from other advanced indus-tries. Thus, the hypercar would have an ultra-light, aerodynamically shaped body fabricated using composite materials. It could utilize a hybrid power train in the form of a small (petrol or perhaps diesel) engine that would power electric motors at the wheels. Lovins argued that such a combination might lead to fuel efficiencies of up to 360 m.p.g. but with the performance of top BMWs. This hypercar concept has been influential outside USA. In the UK it stimulated the Transport Technology Foresight Panel (1995a) to advocate the devel-opment of a 'Foresight vehicle' that is significantly more environ-mentally-friendly than current designs, but which meets mass market expectations of cost, performance and safety. The Panel suggested that a range of motor car power systems could be encompassed within the development programme, including clean fuel IC engines as well as electric and hybrid drives (collectively termed 'low emission vehicles').

V   CONCLUDING REMARKS

Environmental Advocacy versus Critical Analysis

Energy use and the consequent emission of pollutants within indus-trial society are one of the key elements in attempts to improve envi-ronmental quality. Trade-offs between human (economic and social) development and environmental protection are at the heart of a long-term strategy for sustainability. The issues involved are complex, and several cases have been used to illustrate the resulting problems: global warming, nuclear power, and transport and the environment. The solution to green issues are rarely 'black' or 'white', but typically 'grey'. Nuclear power and its perceived environmental impact is often cited, for example, by protagonists and opponents alike to justify their case. It is important that academics, and researchers more widely, approach these issues in a rational manner. The arguments need to be carefully weighed in the balance. There is a great temptation to act as an advocate for some apparently worthwhile, or fashionable, envi-ronmental cause. Paraphrasing John Maynard Keynes: to become a slave of some defunct (or recently popular) environmental theory (or fashion). In the longer term the global environment will be better served by the application of critical analysis, and 'cooler heads'.

The present contribution has argued for the use of an interdisciplinary portfolio of analytical techniques, within a general systems framework. Concepts such as the life-cycle of products and processes, and the importance of clearly defined system boundaries are key elements in environmental problem solving. Some economists would claim that (as a normative discipline) their methods, including cost–benefit analysis, can be extended to incorporate all of society's environmental concerns. In contrast, it has been suggested here that economic techniques may well obscure the impacts of different courses of action, and decision-makers become less well informed rather than the reverse. The latter, particularly if politicians, need to be confronted with the full range of environmental consequences that flow from their decisions. It is too easy to hide behind an aggregate decision criteria, within which the critically important weighing of various impacts are concealed. Nevertheless, CBA can play a useful role as one method within an interdisciplinary toolkit.

Putting environmental issues above everything else misses the point of sustainable development; getting the right balance between the competing demands of both current and future generations over a range of policy areas. Governments need to be open with their population about the choices implicit in weighing these demands, and should highlight rather than obscure the inherent uncertainties. They ought to stimulate debate in the interests of generating a mature, informed and participatory society. Here academics can play an important educational role in the broadest sense; probing the premises on which the 'Doomsday' scenarios are based, and highlighting the uncertainties.

Science, Uncertainty and Climate Change

Discussions about the critical issue of whether climate change on a world scale is due to human activity of simply natural variability has generated rather more heat than light. The protagonists in the global warming debate now seem to have divided into two camps. On one side is the fossil fuel lobby (represented by the US-based Global Climate Coalition) and some economists, such as those associated with the Institute of Economic Affairs. These groups have been campaigning to inhibit action to reduce carbon dioxide emissions pending further research; a position that is clearly in the economic self-interest of conventional energy suppliers. In contrast, the advocates of the enhanced greenhouse effect seem at times to believe in the phenomenon almost literally as an act of (religious) faith. They often react vehemently against any scientific challenge to this belief. Sir John Houghton (1997), for example, who is co-chairman of the IPCC Scientific Assessment Working Group and chairman of the UK's Royal Commission on Environmental Pollution, is quite explicit about the religious motivation behind his environmental concern. In any event, attention has been drawn here to the large scientific uncertainty over anthropogenic climate change, along with alternative hypotheses for global warming (such as the possible link with the 10–12-year solar cycle).

It seems that politicians in all the main political parties in the UK have now accepted the idea of the enhanced greenhouse effect. The danger of this approach is twofold. First, if climate change is not found to be primarily caused by anthropogenic activity (but is simply part of some grand natural cycle), then there is great potential for disillusionment with scientific community among the general population. Secondly, it detracts attention from what might prove to be other, more important, environmental issues, and the need to adopt strategies aimed at securing sustainable development; taking humanity's 'eye off the ball'. A more measured assessment was recently provided by Sir Robert May (1997), the UK Government's Chief Scientific Adviser, who advocated action to reduce carbon dioxide emissions on that basis of what environmentalists term the 'precautionary principle'. Not out of conviction that anthropogenic climate change is currently proven, but because its possible effect over the next century may be damaging and large-scale. Carbon dioxide released into the atmosphere from the burning of fossil fuels is thought to persist for around a hundred years.

The issue of global warming won't go away. At the time of writing (Autumn 1997) it does not appear that European Union will be able to agree the large CO_2 emission reduction targets it would like with non-European OECD member states at the Conference of Parties to the Climate Change Convention in Kyoto, December 1997. In reality Government Ministers can sign up to tight emissions targets for carbon dioxide (perhaps coupled with the introduction of a global system of tradable emission permits, like that championed in the UK by Michael Grubb (1997) of the Royal Institute of International Affairs) in all conscience. This is because, in effect, carbon dioxide acts as a 'marker' for a range of pollutants (SO_2, NO_x and VOCs) that have an impact on the environmental quality of ordinary people on a local and regional scale.

However, Ministers should make their case on wider grounds of sustainability and not rely on currently doubtful arguments about climate change. This will not be easy. In order to reduce pollutant emissions of any kind, it is necessary to tackle both the overall level of resource use as well as the amount of emission per unit of resource consumed. The resource base and so-called 'carrying capacity' of the planet are not well defined; limits are often unclear until they are almost reached. One of the great challenges for the next century is therefore to dramatically improve the efficiency of resource use, particularly of non-renewable energy, across the planet so that humankind will 'tread lightly on the Earth'. Clearly the industrial nations whose societies are by far the most resource intensive, will need to take the lead. A more equitable sharing of world income and resources is likely to be a prerequisite for long-term sustainability. It will require difficult decisions for the West in terms of market intervention to stimulate the development of sustainable technologies, and possibly to induce changes in life-style. These are matters of interregional and intergenerational ethics, rather than purely scientific debate.

Energy and Resource Use Efficiency

Despite the scientific uncertainties over anthropogenic climate change, there is a potential energy strategy that would be robust against any eventuality, and which would satisfy the precautionary principle. This is one based on stimulating improvements in resource use efficiency generally, and the encouragement of energy conservation in particular from the bottom-up. It would involve a consumer-oriented market approach, coupled with intervention by way of a portfolio of measures to counter market deficiencies. These would include economic instruments (such as green taxes devised with the benefit of the experience of some of the EU countries), environmental regulation, and land use planning procedures. Scenarios such as the Factor Four project advocated by Ernst von Weizsaker *et al.* (1997) suggest that economic welfare in the industrial world might be doubled while resource use is halved; thus the Factor 4. Improvements in resource use efficiency of this magnitude would have an enormous knock-on benefit of reducing pollutant emissions that have an impact, actual or potential, on environmental quality. In reality such a strategy requires a major change (paradigm shift) to an energy system that is focused on maximizing the full fuel/energy cycle efficiency, and minimizing the embedded energy in materials and products by way of reuse and recycling. In order to make such an approach a practicable engineering option, it would be necessary to use systems analysis methods to optimize the energy cascade.

Actions taken to reduce pollutant emissions from power stations, road transport, industry, and in the home would have benefits on both a local and global scale. Measures to limit acid rain precursors from electricity generation, for example, will also reduce greenhouse gas emissions. Similarly integrated transport strategies, currently being reviewed by the UK Government, will alleviate congestion in urban centres, reduce toxic vehicle emissions (thereby improving local air quality and health), as well as mitigating the enhanced greenhouse effect. Policies of this type are therefore of a 'win–win' nature. They can be used to focus attention on local and regional environmental concerns to the benefit of ordinary people. Community participation can be encouraged via the Local Agenda 21 process that has been successful in many parts of the UK and elsewhere. It will encourage people to 'think globally, act locally' in a meaningful way. Central government needs to encourage implementation, and develop an enhanced systems modelling capability to ensure that the sum of the parts meet national targets. Thus, the environmental challenges of the new Millennium could be met with a mixture of vision and realism.

VI TOWARDS AN ENVIRONMENT RESEARCH AGENDA

The last Conservative Government in the UK instigated a broad 'technology foresight' exercise in 1994 modelled on similar programmes elsewhere in the OECD. It initially established some 15 independent panels with the task of generating 'visions of the future' that might assist decision-making in both the public and private sectors. This should provide a basis for ensuring that resources are used in support of wealth creation and improving the quality of life. Each panel consisted of members drawn from a wide range of academic, industrial and research backgrounds. They developed networks of business people, engineers and scientists to engage in the foresight process. Individual panels operated in rather different ways with an overall framework, although all sought to identify high priority technological projects that could be implemented (successfully introduced to the market) within a 20-year time horizon. This is what Hammond (1997b) referred to as 'medium-term', while the long-term was considered to stretch out to 2050–2100.

Overlapping energy and environmental issues were examined by three of the Technology Foresight Panels: 'Transport' (1995a), 'Agriculture, Natural Resources and the Environment' (ANRE) (1995b), and 'Energy' (1995c). The recommendations for three thematic projects in the transport area were examined in some detail above. These would all benefit from the sort of critical analysis advocated here. This would most obviously be the case in the proposed 'Foresight Vehicle' initiative, where the more radical suggestions for new vehicles and power units need careful thermodynamic analysis within a systems framework. The ANRE Panel was the only one of those engaged in the environmental sector to lay particular emphasis on monitoring, surveys, process studies, forecasting, and impact evaluation studies. They drew attention to the need for the type of life-cycle analysis and socio-economic modelling recommended in the present work. Obviously the outcome of the Energy Technology Foresight Panel is central to the concerns reviewed here. They identified a number of top priority products/markets within the 20-year timescale, using a Delphi survey of the informed business and technical community. These included: advanced techniques for oil and natural gas exploration and recovery; decommissioning methods for nuclear facilities; clean-coal, combined-cycle and photovoltaic power generation; low emission vehicles; and energy-efficient buildings. In order to support such developments the Panel also recommended the funding of related underpinning technologies and sciences. Beyond the 20-year time horizon the Panel argued for a watching brief to be kept on a number of technologies that might have significant benefit to the UK in the long term: nuclear fusion; advanced nuclear reactors; coal gasification and liquefaction; hydrogen as a fuel; and renewables.

The research agenda identified under the auspices of the Technology Foresight initiative provides a broad framework for future studies in energy and the environment. It had the benefit of drawing on expertise across the industrial and research community. Nevertheless, it is important to examine the impact of the resulting technological and socio-technical developments on society using the critical, interdisciplinary systems approach advocated here. Only in this way will unwanted environmental side effects be minimized, and the challenges of sustainable development be met.

Acknowledgements

The author's research on energy analysis and life-cycle assessment has been supported by research grants awarded by the UK Engineering and Physical Sciences Research Council (most recently under grants GR/J92910, GR/L02227, and GR/L26858). He would also like to acknowledge the support of British Gas plc, who have partially funded his Professorship. However, the views expressed in this chapter are those of the author alone, and do not necessarily reflect the policies of the company. Finally, the author is grateful for the care with which Mrs Heather Golland prepared the typescript and Mrs Gill Green prepared the figures.

Further Reading

G. BOYLE (1996) *Renewable Energy: Power for a Sustainable Future*, Oxford, Oxford University Press.

N. CALDER (1997) *The Manic Sun: Weather Theories Confounded*, London, Pilkington Press.

R. EDEN, M. POSNER, R. BENDING, E. CROUCH and J. STANISLAW (1981) *Energy Economics: Growth, Resources and Policies*, Cambridge, Cambridge University Press.

J. GOLDEMBERG (1996) *Energy, Environment and Development*, London, Earthscan Publications.

T. E. GRAEDEL and B. R. ALLENBY (1995) *Industrial Ecology*, Englewood Cliffs, NJ, Prentice-Hall.

J. HOUGHTON (1997) *Global Warming: the Complete Briefing*, 2nd edn, Cambridge, Cambridge University Press.

N. DE NEVERS (1995) *Air Pollution Control Engineering*, New York, McGraw-Hill.

J. RAMAGE (1983) *Energy: A Guidebook*, Oxford, Oxford University Press.

M. SLESSER (1978) *Energy in the Economy*, London, Macmillan Press.

D. E. WINTERBONE (1997) *Advanced Thermodynamics for Engineers*, London, Arnold.

References

M. A. ANDREWS (1991) *The Birth of Europe*, London, BBC Books.

Automobile Association (1997) 'Tracking emissions from UK vehicle exhausts: the AA toxic tailpipe indices', Basingstoke, AA.

R. U. AYRES (1995) 'Life cycle analysis: a critique', *Resources, Conservation, and Recycling*, 14, pp. 199–223.

BP Statistical Review of World Energy (1997), June (annual).

British Governmental Panel on Sustainable Development (1997), Third Report, Jan.

L. W. CANTER (1996) *Environmental Impact Assessment*, 2nd edn, New York, McGraw-Hill.

D. CARSON (1997) 'Global warming: the basis for concern', Annual Joint Lecture, University of Bath, 13 Mar. (unpublished).

P. CHAPMAN (1976) 'Methods of energy analysis', In Blair, I. M., Jones, B. D. and Van Horn, A. J. (eds), *Aspects of Energy Conversion*, Oxford, Pergamon.

J. G. COLLIER (no date) 'Nuclear power—clean energy for the 21st century', Nuclear Electric plc.

Commission of the European Communities (1990) 'Energy for a new century: the European perspective', *Energy in Europe* (special issue), July.

N. DA COSTA, G. P. HAMMOND and G. RUCH (in preparation) 'Life-cycle energy analysis of electric and parallel hybrid powered passenger cars'.

Department of Trade and Industry (DTI) (1996) *Digest of United Kingdom Energy Statistics 1996*, London, HMSO.

R. DORFMAN and N. S. DORFMAN (eds) (1993) *Economics of the Environment*, 3rd edn, New York, W.W. Norton.

J. H. FREMLIN (1987) *Power Production: What are the Risks?*, Oxford, Oxford University Press.

E. FRIIS-CHRISTENSEN and K. LASSEN (1991) 'Length of the solar cycle: an indicator of solar activity closely associated with climate', *Science*, 254, pp. 698–700.

W. VAN GOOL (1997) 'Energy policy: fairy tales and factualities', In Soares, O. D. D. *et al.* (eds), *Innovation and Technology: Strategies and Policies*, Dordrecht, Kluwer.

M. P. GOVER, S. A. COLLINGS, G. S. HITCHCOCK, D. P. MOON and G. T. WILKINS (1996) *Alternative Road Transport Fuels—A Preliminary Lifecycle Study for the UK*, 2 Volumes, London, HMSO.

Greenpeace (1994) *No Case for a Special Case: Nuclear Power and Government Energy Policy*, London, Greenpeace.

M. GRUBB (1991) 'Buying a better climate', *Prospect*, Aug./Sept., pp. 58–61.

The Hadley Centre (1995) *Modelling Climate Change 1860–2050*, Bracknell, The Meterological Office.

G. P. HAMMOND (1997a) 'Energy forecasting: visions of the future, lessons from the past', IMechE Seminar: The Competitive Market for Energy—1998 and Beyond, IMechE, London, 8 Apr.

G. P. HAMMOND (1997b) 'Nuclear energy into the 21st century', In Soares, O. D. D., Martins da Cruz, A., Costa Pereira, G., Soares, I. M. R. T. and Reis, A. J. P. S. (eds), *Innovation and Technology: Strategies and Policies*, Dordrecht, Kluwer.

G. P. HAMMOND (1998) 'Alternative energy strategies for the United Kingdom revisited: market competition and sustainability', *Technological Forecasting and Social Change*, 59, pp. 131–51.

G. P. HAMMOND and R. M. MACKAY (1993) 'Projections of UK Oil and Gas Supply and Demand to 2010', *Applied Energy*, 44, pp. 93–112.

J. T. HOUGHTON *et al.* (eds) (1996) *Climate Change 1995: the Science of Climate Change*, Cambridge, Cambrdge University Press.

P. M. KELLY and T. M. L. WIGLEY (1992) 'Solar cycle length, greenhouse forcing and global climate', *Nature*, 360, pp. 328–30.

R. A. KERR (1996) 'A new dawn for sun-climate links?', *Science*, 271, pp. 1360–1.

G. LEACH, C. LEWIS, F. ROMIG, A. VAN BUREN and G. FOLEY (1979) *A Low Energy Strategy for the United Kingdom*, London, IIED/Science Reviews.

J. LEAN (1997) 'The sun's variable radiation and its relevance for Earth', *Ann. Rev. Astron. Astr.*, 35, pp. 33–67.

J. J. LEE, P. O'CALLAGHAN and D. ALLEN (1995) 'Critical review of life cycle analysis and assessment techniques and their application to commercial activities', *Resources, Conservation, and Recycling*, 13, pp. 37–56.

A. B. LOVINS (1977) *Soft Energy Paths*, Harmondsworth, Penguin.

A. B. LOVINS, J. BARNETT and L. H. LOVINS (1993) *Supercars: the Coming Light-vehicle Revolution*, Snowmass, Rocky Mountain Institute.

D. MADDISON, D. PEARCE, O. JOHANSSON, E. CALTHROP, T. LITMAN and E. VERHOEF (1996) *Blueprint 5: the True Cost of Road Transport*, London, Earthscan.

R. MAY (1997) *Climate Change*, London, Office of Science and Technology.

D. L. MUNBY (1976) 'Economics of resource use'. In Blair, I. M., Jones, B. D. and Van Horn, A. J. (eds), *Aspects of Energy Conversion*, Oxford, Pergamon.

Nuclear Electric plc (1994) Submission to the UK Government Review of Nuclear Energy, Volume 2. *The Environmental and Strategic Benefits of Nuclear Power*, Barnwood, Nuclear Electric.

F. ROBERTS (1978) 'The aims, methods and uses of energy accounting', *Applied Energy*, 4, pp. 199–217.

Royal Commission on Environmental Pollution (RCEP) (1994) *Transport and the Environment, 18th Report, Cmnd 2674*, London, HMSO.

Royal Commission on Environmental Pollution (1997) *Transport and the Environment—Developments Since 1994, 20th Report, Cmnd 3752*, London, SO.

H. SVENSMARK and E. FRII-CHRISTENSEN (1997) 'Variation of cosmic ray flux and global cloud coverage—a missing link in solar-climate relationships', *J. Atmos. Solar-terr. Phys.*, 59, pp. 1225–32.

J. SZARGUT, D. R. MORRIS and F. R. STEWARD (1988) *Exergy Analysis of Thermal, Chemical, and Metallurgical Processes*, New York, Hemisphere.

Technology Foresight (1995a) *Progress Through Partnership 5: Transport*, London, HMSO.

Technology Foresight (1995b) *Progress Through Partnership 11: Agriculture, Natural Resources and the Environment*, London, HMSO.

Technology Foresight (1995c) *Progress Through Partnership 13: Energy*, London, HMSO.

G. WALL (1987) 'Exergy conversion in the Swedish Society', *Resources and Energy*, 9, pp. 55–73.

G. WALL (1990) 'Exergy conversion in the Japanese society', *Energy*, 15, pp. 435–44.

E. VON WEIZSACKER, A. B. LOVINS and L. H. LOVINS (1997) *Factor Four: Doubling Wealth, Halving Resource Use*, London, Earthscan.

World Commission on Environment and Development (1987) *Our Common Future, The Brundtland Report*, Oxford, Oxford University Press.

World Energy Council (1993) *Energy for Tomorrow's World*, New York, St Martin's Press.

8 Finding Poisons: Techniques for Gross Pollution Monitoring

William Bains

Summary

The demands for tighter regulation of waste processing outflows and for more intensive methods for processing many industrial wastes have combined to build a demand for methods of monitoring a wide a range of pollutants in waste water. Such 'Gross Pollution Monitor' (GPM) systems must be able to detect many types of toxin. Whereas chemical sensors are limited to detecting one class of chemical compound, biologically based sensors can in principle detect any potential pollutant. This chapter describes a new GPM method, the Metabolic Ultra Violet Sensor (MUVS) method, its scientific background and initial technical development. In addition, the reasons why an industry which needs techniques such as MUVS cannot fund its development are discussed, and whether this is likely to change.

I INTRODUCTION

Background

One of the topics of an environmental research agenda must be methods for determining when our environment is degraded. This chapter will address one aspect of this; detectors for general or gross pollution, and their use as environmental alarm systems.

Detecting toxins in water is a pressing commercial problem (Bains, 1992, 1993c). Their use in protecting water quality is clear, both monitoring industrial intake water for quality and monitoring outflow for regulatory compliance. They also have purely pragmatic uses in protecting sewage treatment works, which are in effect large mixed fauna fermenters, from the effects of toxins that kill those fauna. This second application is the primary market for toxin 'alarm' systems, as they are for the protection of an industrial process, and not 'merely' required for regulatory compliance, and failure of that protection has substantial economic impact in plant down time and cleaning costs if an overload of a toxic material were to be allowed entry. Economic drivers of this sort make funding development of such alarm sensors much more attractive.

Toxin alarms also share technology with the search for drugs which are selectively toxic to pathogens or to cancer cells, and this will be briefly discussed below.

The Technical Problems of Toxin Alarm Systems

Toxic monitors are of two types—laboratory tests for specific toxins, and rapid 'alarm' systems for detecting *any* toxin. The former must be chemically exact, but can be slow. The latter can be chemically inexact, but must be fast. This second is a chemically difficult problem to solve, as many different agents can be toxic for many reasons, and we want to be able to detect all of them. In fact, we want to detect a biological effect, and so this problem lends itself to a biosensor.

The 'best practice' system for general toxicity monitoring is the 'trout monitor', which uses swimming or breathing movement of trout as a monitor for the quality of the water in which they are swimming. Avoidance or alarm swimming by the trout is taken as a sign that the water they are in is contaminated. This is a biosensor, using the trout as a detector mechanism. It has substantial practical problems with reliability, as many non-toxic events can also 'alarm' the trout, and so there are a range of alternatives using simpler organisms. These are listed in Table 8.1. Some of these are commercially available, but all suffer from substantial technical drawbacks. Most use micro-organisms as surrogates for humans, and thus they are inherently unable to detect neurotoxins. Most also lack photosynthetic apparatus, and so cannot detect pesticides such as DCCD.

The principle difference between the microbial systems is the method of readout. Most use bulk batch culture, and a gas sensor output. These suffer from fouling of the sensor, and so systems which do not require gas sensor readouts have been developed to use luminescence or direct electrochemical coupling of the bacterial electron transport chain to a suitable electrode as a direct measure of bacterial energy metabolism.

The MUVS sensor uses an indirect but not gas-related measuring

Table 8.1 Microbial gross pollution monitor systems

System name	Manufacturer/developer	Mode of action
Trout Monitor	Various—originally validated by Water Research Council, UK	Measures trout swimming, breathing or avoidance behaviour
Microtox	Microbics Corp., Carlsbad, CA	Measures light output from luminescent bacteria—light declines on poisoning
ToxAlarm	Genossenschaft Berliner Ingieurcollective, Berlin, Germany	Measures O_2 uptake (respiration) in batch culture of bacteria
RodTox	Kelma, Niel, Belgium	Measures O_2 uptake (respiration) in batch culture of bacteria
Micro OxyMax	Columbus Instruments Intl. Corp., Columbus, Ohio, USA	Measures O_2 uptake and CO_2 generation (respiration) in batch culture of bacteria
BOD 2000	Central Kagaku Corporation, Tokyo, Japan	Measures CO_2 generation (respiration) in batch culture of bacteria
Products in development		
	Lange	Flow cell system to measures CO_2 generation (respiration) semi-continuously
	A.P.F. Turner, Cranfield Biotechnology Ltd, Cranfield, UK	Direct electrochemical coupling of photosynthetic bacteria to electrodes (measures bacterial respiration)
	Rawson *et al.*, Luton College (now Luton University), Luton, UK	Direct electrochemical coupling of bacteria to electrodes (measures bacterial respiration) (measures bacterial respiration)
MUVS	Bains, PA Consulting Group, Melbourn, Herts, UK	UV spectrometry—measures respiration status

system. It has the advantage that it is quite general—it does not require optimization for each new bacterium.

II THE TECHNICAL RESEARCH PROGRAMME

The MUVS Effect

Figure 8.1 illustrates the basic MUVS effect. Whole living bacteria absorb ultraviolet light strongly in the region 190–300 nm, with a characteristic spectrum. As the bacteria is poisoned, it can be observed that the absorbance in most of this region is unaltered (as you would expect, as poisoning is relatively subtle), but the peak at between

Figure 8.1   *The MUVS effect*

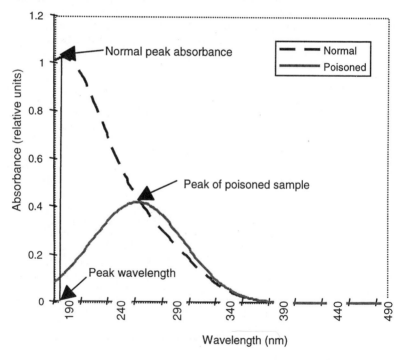

Note: Shown are two absorbance spectra, for bacterial suspensions in water from 190 nm to 500 nm, for bacteria poisoned with sodium azide and for unpoisoned, healthy bacteria. Spectral features used to measures changes quantitatively are shown.

190 nm and 200 nm drops dramatically, ultimately disappearing completely. This appears to be quite general, and is very easy to measure: Table 8.2 lists the organisms for which a 200 nm-peak 'poisoning' effect has been confirmed. It also works for a wide range of toxins (listed in Table 8.3: only a few of the very large number of combinations of toxin and organism have been tried).

The effect is also 'dose-dependent', i.e. the more the organism is poisoned (as measured by its reduction in growth), the more the peak height is reduced. Figure 8.2 shows an example of such a dose response curve as measured by MUVS, and Figure 8.3 the dose response curve for the same organism:toxin combination as measured by the failure of the bacteria to grow (the standard test method). Both

Figure 8.2   *Dose—response curve for MUVS*

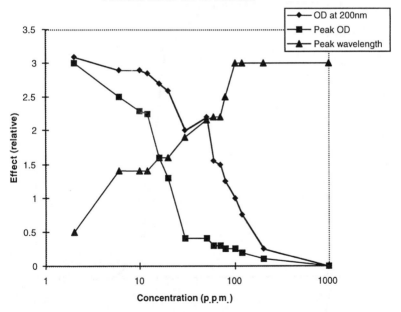

Concentration-dependence of MUVS effect of sodium azide on *B. subtilis*

Note: Dose response curves for MUVS. Shown are three measures of the change in the mid-UV spectrum: OD (absorbance) at 200 nm, the wavelength of peak absorbance between 190 nm and 260 nm, and the absorbance at that peak.

Table 8.2   Organisms for which the MUVS effect
(selective, dose-dependent reduction in the
absorbance at 190–200 nm of whole living cells)
has been experimentally confirmed

Bacillus subtilis
E. coli
Mammalian epidermal fibroblasts
Pseudomonas aeruginosa
Saccharomyces cereviseae
Sarcina lutea
Tetrahymena pyriformis

Figure 8.3   *Dose–response curve for growth inhibition*

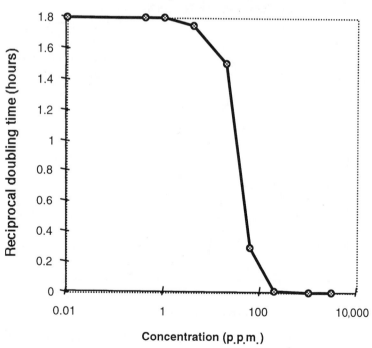

Growth inhibition of *Bacillus subtilis* by sodium azide

Note: Growth rate (reciprocal doubling time) of Bacillus subtilis cultures grown with differing concentrations of sodium azide.

Table 8.3 Toxins (by class)   185

Compound	Class	Mode of action	Concentration of toxin giving half maximal response (mM)
2-mercaptoethanol	General toxin	General protein-unfolding agent	1.5–1.9
Cadmium chloride	Heavy metal	Broadly toxic	0.4–1.6
Chloramphenicol	Antibiotic	Inhibits protein synthesis	0.15–0.24
Copper-I chloride	Heavy metal	Oxidatively toxic	0.1
Copper-II sulphate	Heavy metal	Broadly toxic	1.6
CTAB	Biocide	Disrupts cell membrane	0.22–0.24
Cycloserine	Antimetabolite	Inhibits cell wall synthesis	0.39–0.78
Iodoacetic acid	Antimetabolite	Inhibits central energy metabolism	0.2–0.26
Kanamycin	Antibiotic	Inhibits protein synthesis	0.32–0.34
Mercury-II chloride	Heavy metal	Broadly toxic	0.34–0.68
Nalidixic acid	Antibiotic	Inhibits DNA synthesis	0.08–0.17
Nickel chloride	Heavy metal	Broadly toxic	0.07–0.13
Penicillin-G[1]	Antibiotic	Inhibits cell wall synthesis	
Phenol	Biocide	Broadly toxic protein unfolding agent	0.0005–0.0015
Rifamycin	Antibiotic	Inhibits protein synthesis	0.085–0.103
Rotenone	Antimetabolite	Inhibits energy metabolism	0.38–0.61
Silver nitrate	Heavy metal	Oxidatively toxic	0.13–0.2
Sodium arsenate	Heavy metal	Broadly toxic	0.033–0.1
Sodium azide	Biocide	Potent broad-spectrum biocide	0.23–0.46
Sodium bisulphate	Disinfectant	General oxidative disinfectant	0.86–0.96
Sodium cyanide[2]	Biocide	Inhibits energy metabolism	51–200
Sulphamethazine	Antibiotic	Inhibits central metabolism	0.04–0.07
Tween-20	Disinfectant	Detergent—disrupts cell membrane	10–15
Trimethoprim	Antibiotic	Inhibits central metabolism	0.09–0.2
Zinc sulphate	Heavy metal	Broadly toxic	0.06–0.11

Notes: [1] Insensitive strain; [2] MUVS is inherently insensitive to cyanide. For discussion, see Bains, 1993b.

methods show a LD50 of around 10–20 p.p.m. of sodium azide in *Bacillus subtilis*, showing that the MUVS measurement mirrors the 'real' toxic effect of this poison. The exact 'lethal dose' shown by MUVS depends on how the spectrum is measured, a critical point for industrial implementation.

This was a quite fortuitous observation, discovered by chance, and the mechanism is still unclear. A range of observations suggest that the absorbing moiety is reduced haemes in the cytochromes of the oxidative phosphorylation chain (Bains, 1993b). However it works, it fulfils the criteria for a technology for a gross pollution monitor system—it is fast, easy to adapt, works on many types of toxin and does not require a gas electrode to be dipped into a river sample.

After these proof of principle experiments, we approached Anglian Water to sponsor a trial of the technology to see if it would be appropriate for their use. This lead to two series of further experiments.

Applicability to Environmentally Important Toxins

Table 8.4 summarizes an experiment looking at some combinations of toxins and detecting organisms, and compares the limits of detection to those reported for the Trout Monitor. These were a range of test pollutants selected by Anglian Water, with which they had had pollution incidents in the past. Generally, the MUVS test is as good as (or slightly better than) the Trout Monitor. However, for some MUVS was very much worse, notably cadmium. Why the bacteria are so resistant to cadmium is still unexplained.

Reliability of the Test

Reliability is a critical factor in choosing an environmental monitoring system—the system must work in the field for months without supervision, and give a low rate of false positive or negative results. Figures 8.4 and 8.5 summarize the result of the reliability of the MUVS technology. In summary, zinc and phenol were chosen as two environmentally relevant toxins for which the MUVS result was consistent and comparable to the Trout Monitor. Three real river water samples were spiked (provided by Anglian Water) and a sample of distilled water (DW) with three concentrations of those toxins—low (the lowest that we thought MUVS would detect), medium (the level Anglian water wanted detected) and high (a concentration we were confident that both Trout Monitor and MUVS could detect), and

Table 8.4 Preliminary observations on MUVS detection of
environmentally relevant toxins. The table shows the minimum sensitivity
of the MUVS test to different toxins (concentrations in ppm).
Also given in minimum sensitivity of the Trout Monitor system
(data from Anglian Water)

Toxin	Trout monitor	MUVS: test organisms				
		B. subtilis	P. aeruginosa	S. lutea	S. cereviseae	T. pyriformis
Sodium arsenate	–	–	15	5	20	5
Cadmium chloride	0.025	–	150	125	300	75
Copper-II sulphate	0.06	10	–	–	–	–
Chromium (IV) sulphate	2–25	2	–	–	–	–
Mercury (II) chloride	–	–	10	10	20	5
Nickel sulphate	–	–	20	10	15	10
Pentachloro phenol	0.24	–	–	1	–	1
Phenol	4	–	1	0.5–1	1.5	0.5
Zinc sulphate	40	–	20	10	20	20

tested them with two organisms using the MUVS method on 10 consecutive days.

The data is summarized in Figures 8.4 and 8.5, which show the
results of comparing the spectrum of one bacterium poisoned with
one toxin on one day with the *same* bacterium poisoned in the *same*
water sample, but with no toxin, on another day. Shown here is the
level of statistical significance in two different spectral measures.
Anglian Water was consulted extensively about what constituted
'success' for a test of this sort. Having a difference significant at 4–6
standard deviations was considered success for a 'pass/fail' test, i.e.
one in which a toxin could be detected or not. A difference of at least
8 standard deviations was needed for a quantitative test, i.e. one which
could reliably indicate *how much* toxin was there. MUVS passed the
pass/fail criterion for zinc and for phenol. Performance at the more
demanding quantitative scale was good for phenol, variable for zinc.

This study shows that the assay works, it can detect toxins, and can
do so fairly reliably.

Figure 8.4   *Reproducibility of MUVS effect: Comparison of absorbance at 191 nm and 200 nm*

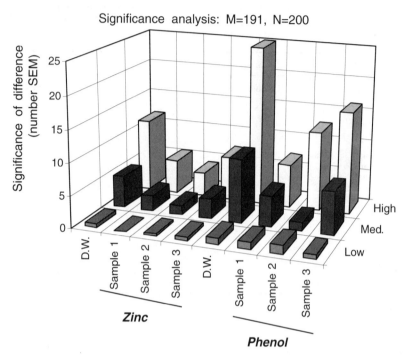

Note: Two representative plots of the significance of the MUVS effect are shown. In summary, the bar heights (truncated at 25) show the significance of the difference between the 'no toxin' signal and the 'toxin' signal for different concentrations of toxin and different water samples, tested over 10 runs. Figures 8.4 and 8.5 show different methods of summarising the MUVS signal.

Flow Cell Sensor System

Because this is a batch system, the possibility of using this methodology was examined in order to build a flow cell version of the test. The details of this have been published in Bains (1994). The apparatus was very simple—the gel was standard laboratory agarose, held in a clip made from a bent paper clip and wedged into the quartz cuvette using cut-down cork (which leaked). However, surprisingly, it worked first time, and showed that water could be run past such a sensor for at least hours with no substantial loss of function. A 'dipstick' version of this gel was displayed for the Toshiba Year of Invention award, but it

Figure 8.5   *Reproducibility of MUVS effect: Comparison of absorbance at 193 nm and the peak between 190 nm and 250 nm*

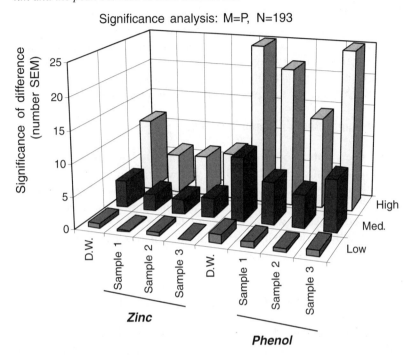

did not work very well, being made out of balsa wood. The system is also quite impractical for field use. However, it demonstrates that a flow cell format can be developed using the MUVS technology.

Other Applications: Antibiotic Susceptibility Testing

As well as sensing general toxins, MUVS can detect specific bactericidal agents such as antibiotics (see Table 8.3). Working in reverse, can MUVS see if an organism is susceptible to a known bactericidal agent, i.e. can it find if an organism is susceptible to an antibiotic (a task of substantial medical relevance)? A short study showed that it could. Several different organisms with differing susceptibility to Penicillin G were tested for (a) whether they grow in the presence of a range of concentrations of the drug and (b) whether that concentration gives a MUVS effect. The concordance in this admittedly crude experiment is 100%.

Other Applications: Consortia of Cells

Can MUVS work if several cells are mixed? It can. MUVS was run using a consortium of organisms from a local pond—mostly protozoa. The consortium actually detected sodium azide rather more sensitively than laboratory organisms. This means that the consortium of organisms that are actually present in a sewage works could be used in a MUVS test for things that would poison that works.

Other Applications: Mammalian Cells

It has been hinted above that mammalian cells can also give a MUVS response. Because the laboratory I was working in was not really set up for mammalian cell culture a short experiment was performed on cultured skin fibroblasts. It was shown that they showed changes in their absorbance between 190 nm and 240 nm when stressed by hypotonic solutions or environmental insult (in this case an excess of trypsin).

III COMMERCIAL RESPONSE TO MUVS TECHNOLOGY

I have been involved primarily in the technical development of MUVS, so the observations on the business success or otherwise of the technology are quite anecdotal.

The commercial response to MUVS has been disappointing. On the environmental applications, Anglian Water was very encouraging in the research programme, recognizing that this was a potential technology for replacing a system that they need, but which does not work. They committed substantial resources to this, but when the initial proof of concept had been finished, they did not want to take it forward to development of a system that could be placed in the field, for three reasons.

(1) Even with an aggressive development plan it would be 7 years before this could be placed in use as a validated field system, and that was too long a timescale. The reasoning in most industries is that waiting 7 years for solutions is too long—they need solutions to today's problems today—so they will not waste resources working on tomorrow's solutions. This begs the question 'where will tomorrow's solutions come from if not today?':

the answer is generally perceived to be 'research' in some vague, semi-academic context, where it is perceived at all. (This is not true of the pharmaceutical industry, and some sectors of the computer industry, both of which support very speculative and long-term research on occasion.)

(2) The market sizes are too small to support the development. We performed a rough estimate of the total market size for a gross pollution monitor in Europe, based on knowledge of the number of water sites tested by some UK water authorities, the number of severe pollution-hotspots which polluting industries might test, and the number which might be sold to testing laboratories. The result showed a market for about 1,000 systems of £10,000 each (a moderate cost for such a system), or twice that if the machine costs half as much—in both cases, the overall market size is about £10 million.

(3) This makes developing the system for sale unattractive. It is just about enough to cover the cost of development, manufacture and distribution, without any after-sales support or repair (which is bound to be necessary for the first systems, no matter how well they are developed.) Celsis, an environmental testing biotechnology company, echoed this view, as did ABB-Kent-Taylor and several smaller equipment companies. Although everyone agrees that it would be economically justifiable to *buy* a system, it is not worthwhile to *develop* one. This applies whether the product is to be used in-house or sold to third parties. Even the public relations value inherent in such a programme appears to be insufficient to tip the scales. To an extent, MUVS would be adverse publicity, as it might find problems which were previously unknown, and for which they can then be criticized.

It should be pointed out that MUVS comes out quite well on both these arguments. All other systems listed in Table 8.1 are inherently more complicated, and hence will inherently take longer and cost more to develop and more to manufacture. So all arguments apply to *any* environmental detection system. While there may be a need now, there is not an economic path to fulfilling that need in 2002.

IV FURTHER DEVELOPMENT PROGRAMME

The timescale for development is long because the programme:

- must develop the test to a point where it can be automated completely reliably, operating in a sewage works under a wide range of conditions for weeks on end without supervision;
- perform comparably not only to the trout monitor but also to conventional chemical assays over a period of at least a year in the field;
- be engineered in such a way that the consumable element (the bacteria, and any other reagents) can be made, stored for months, transported and used with better than 99% reliability.

When faced with the uncertainties in doing this, the water industry not unreasonably states that it would rather invest in conventional technology because, although it may be more expensive, less reliable and less effective, at least it can be relied upon in the field.

One path then is to develop the technology using someone else's money, and then apply it to the water industry. The healthcare industry has substantially more money than the water industry, and invests a far larger proportion of it in R&D, so the former was approached on two fronts. Antibiotic susceptibility testing is an established diagnostic tool, and for some applications (such as finding out which antibiotic to use on a patient dying of septicaemia), timing is critical. MUVS offered a faster way of doing antibiotic susceptibility testing. Discussions with Unipath, the major supplier of antibiotic susceptibility test equipment in the UK, were encouraging. They wanted to upgrade their product range substantially, and were looking for new, patented and faster technology to do so. But MUVS was too unproved, and an industry recession meant there was no funding for speculative development.

The other healthcare application was to use the test as a drug discovery tool, to find whether potential antibiotics actually killed bacteria, The advantage of the technology was that exactly the same system could be used on mammalian cells to show that the chemicals did *not* kill humans. Despite talking to five leading, research-based, profitable pharmaceutical companies, there was no to interest generated. This was partly because the trend in this very fashion-conscious field had swung away from whole-cell assays to isolated biochemical components, but partly because they could not see how this would fit in with a highly automated 96-well plate format of drug screening, which is the paradigm in the field and in which all major drug companies have together invested about £1 billion in automation.

Venture capital companies were approached with a view to setting up MUVS as an independent company, to develop it to the point where major corporations would become interested. However, the technology was felt to be at too early a stage, even for those venture capital groups which aim at very early stage companies.

As a research project, even one in a commercial setting like PA Consulting Group, the MUVS programme has reached proof of principle that the basic science works. But it is a long way from being a candidate development project. There is a tremendous amount of 'applied research' to do to confirm that it is sensitive enough, reliable enough, robust enough, etc. to be industrialized. Until that has been done, it is a high-risk research project for industry. And, unlike the research on the basic MUVS effect (which is generic) the applied research is specific for specific applications (soil leach tests, ground water, waster water, clean water, water disinfectants) and specific requirements (practical use, legislative requirement, commercial vs individual users) and so on. That systematic exploration of the technology costs millions of pounds and takes years, and a research operation cannot support that. So there is a gap between research and development.

In discussion, the meeting asked whether this would have gone better if it had been a collaborative venture with the eventual end-users. Systematic research suggests that such projects are much more likely to succeed than ones where a technical solution is developed in isolation from the people who actually have the problem. This is undoubtedly true (if the end-users can decide what their problems are), but it is not a fair comparison. This was not a case of discussing a problem with (in this instance) a water company, and then working together to a solution. The MUVS programme is more akin to scientific discovery. Here is an effect which has potential value as an environmental tool, and possibly as a tool in other industries. As a discovery it cannot be made 'to order' as part of a pre-existing collaboration. It is a model for the scientist who says 'I have something really neat. What can I do with it?'.

In any case, the initial discovery of the MUVS effect was driven by the known need of one of the world's largest water processing equipment companies for a GPM system. The work described here was largely done in collaboration with Anglian Water, who were far-sighted enough to invest their time, money and effort in working with us to explore how much work would be needed to get MUVS into the field. The samples of river water, for example, were provided by them, as were test protocols, statistical methods, and the toxins they thought we should test. So much of the work on the environmental aspects of the sensor described here was done in close collaboration with end users.

V   CONCLUSIONS—NEXT STEPS

The generality of the MUVS observations has been established, and some preliminary data shows that it could be applied to several industries. Despite this, none of them is interested. In May 1996 PA Consulting Group, the original owner of the patents, themselves lost interest in MUVS, because it could not be 'sold'. MUVS had reached the gap between academic research (which is relatively cheap, but brain- and time-intensive) and commercial development (which is more dependent on money). There is a major dearth of funding for transferring technology over this gap in the UK, and neither venture capital nor companies in the field exist to bridge the gap. The MUVS project falls exactly in the middle of this. As a research programme MUVS is finished—the results are collected, the papers published. As an industrial development project it is not ready to start. Only an external event, such as the 'Camelford' water pollution, could provide the incentive for industrial development of MUVS at this stage. (MUVS would have detected the aluminium released at Camelford.)

This leaves a range of research in an academic context which might both develop interesting pieces of fundamental knowledge and could advance the technology to a point where even British industry can believe in it. Much of this revolves around what exactly MUVS is, what it is detecting, and why it appears such a general indicator of toxicity. This is not 'development' in an industrial sense, but will lend credence to the eventual application of MUVS in industry.

Postscript

Since this talk I have left PA Consulting Group where MUVS was developed, and joined Merlin Ventures, a venture capital company set up specifically to bridge the gap between academic research and development company. Since inception in March 1996 Merlin has funded five companies (August 1997), although none yet in environmental technology. None of has have a 'product', and four are based solely on academic research results and the expertise that developed them.

Meanwhile, opinion in the water industry is moving towards recognizing the value of 'toxicity-based consent' (i.e. allowing users to discharge water into the environment based on its global toxicity rather than on specific chemical parameters), raising the possibility that MUVS may become more commercially attractive in the future.

References and Further Reading

W. BAINS (1992) 'Sensors for a clean environment', *Bio/Technology*, 10, pp. 515–18.

W. BAINS (1993b) 'Detection of cellular metabolic state using mid-ultraviolet spectrometry', *Chimica Oggi,* 11:(10), PP. 49–53.

W. BAINS (1993c). 'Microbial biosensors as environmental alarm systems', *International Environmental Technology Guide*, pub Int. Labmate Ltd, pp. 35–7.

W. BAINS (1994) 'A spectroscopically interrogated flow-through type toxicity biosensor', *Biosens. Bioelectr.*, 9, pp. 111–17.

9 Controlling Emissions

Stuart A. MacGregor

Summary

As pollution issues become more important, the emissions from motor vehicles have been brought into sharp focus. Increasingly stringent emissions legislation has highlighted the need for improved technology in motor vehicles. This article reviews the current trends in legislation and considers the basic mechanisms for pollutant prevention. Advances and future developments in internal combustion engine technology are also considered.

I INTRODUCTION

Ask anybody what really concerns them about the environment and the answer will probably be emissions or energy consumption. The two are very closely linked together and it is difficult to consider one while neglecting the other. The motor car will probably be cited as the biggest culprit, some concern may be expressed about power generation, but nobody is prepared to give up the freedom which, they consider that the motor car offers them. They will express concerns about the emissions from power production and the potential problems of nuclear power especially post-Chernobyl. Other sources may be cited but the road vehicle is something we seem very reluctant to give up.

The first concerns about emissions were expressed in the mid-1960s. This was especially so in California, in Los Angeles the atmospheric conditions led to the formation of photochemical smog. This was due to the NO_x and HC in the exhaust gases from the many thousands of cars using the roads. It comes as no surprise that America has the strictest emission laws in the world, with the legislation in California driving the whole programme. The rest of the world is now falling into line as regards the emissions standards now being set.

The trends in emissions can be seen in Figure 9.1. This highlights some of the milestones along the way to reduced emissions and

Figure 9.1   *Pollutant emissions since 1964*

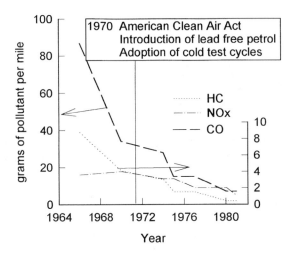

exhaust gas legislation. Prior to 1966 there was no exhaust emission control. This state of affairs continued until about 1970 when the need for legislation was recognized. At that time the American Clean Air Act came into being. While this was happening there were a number of other developments coming along, e.g. the introduction of lead free petrol was introduced. This had two effects, the first was to make people aware of the need for the reduction of emissions. However, there was also an adverse effect with the introduction of lead free petrol, it made people complacent with the feeling that 'I've done my bit for emissions, so what's everybody else doing'. In addition to the introduction of lead free petrol, the problem of sensible test procedures on which emission performance could be measured were considered, with the adoption of more realistic test cycles. Figure 9.2 shows a typical example of a drive cycle. This represents a much better method of emissions testing, compared with steady state tests, as it introduces the effects of warm up and transients which are almost certain to influence the production of pollutant emissions.

During the early part of the 1970s the possibility of reducing emissions by further engine developments was being considered. This included variations in ignition timing, thermal reactors and exhaust gas recirculation (EGR). These topics will be considered in more detail later. As the technology advanced, the possibility of exhaust gas treatment in the form of catalytic converters became a real possibil-

Figure 9.2   *Typical drive cycle*

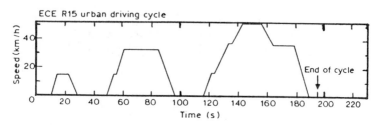

The ECE R15 urban driving cycle; the cycle is repeated four times and is preceded by a warm-up idle time of 40 s (total test duration is 820 s)

ity. These developments continued until now they are accepted as an essential part of every new car.

In Figures 9.3 and 9.4 the picture from more recent times can be seen with the type of emission legislation that is required now. It also gives some indications as to the future trends that are likely to be introduced. Figure 9.3 details the situation in America, which reflects the trend to the development of diesel engines for heavy-duty low speed applications. Figure 9.4 reflects the coming trends in the European Community. The small high-speed diesel is becoming increasingly popular as a power unit for the passenger car market. One has only to look at any car showroom to see that most manufacturers offer small diesel cars today. The emission regulations apply to land vehicles but it will not be long before emission regulations are applied to all forms of transport. The International Maritime Organization, part of the UN, is in the process of imposing emission regulations to which all ships will be subject.

II   BACKGROUND

Having considered the development of emission legislation, it is now useful to look at where the main problem of exhaust emissions lies. The main emissions for a spark ignition engine are carbon monoxide (CO), oxides of nitrogen (NO_x) and hydrocarbons (HC). The levels of these emissions will vary from one engine to another. They are found to be dependent on a number of factors, such as:

Figure 9.3   *Emissions regulations in USA*

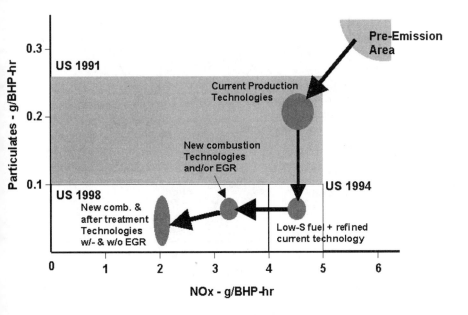

- ignition timing
- load
- speed
- air fuel ratio
- fuel composition

The variation of exhaust gas emissions with air fuel ratio is shown in Figure 9.5. In the case of diesel engines we should add particulate and noise to the list of pollutants given above. Carbon monoxide is most concentrated in fuel rich mixtures, due to incomplete combustion. In lean mixtures CO is still found due to dissociation. By operating with excess air (lean burn), it is possible to significantly reduce the concentration of hydrocarbons in the exhaust gases. However, it is possible to increase the level of excess air to the point where the reduced flammability of the mixture causes problems and results in an increase in HC. Finally, the production of NO_x is a strong function of flame temperature; high flame temperatures will result in high levels of NO_x.

The diagram in Figure 9.6 shows the main sources of emissions for

Figure 9.4   *Emissions regulation in European Community*

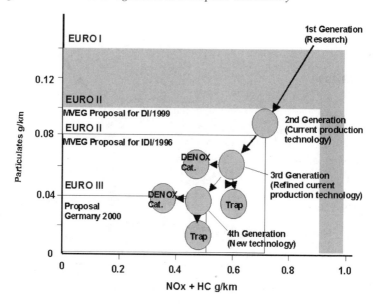

Figure 9.5   *Variation of pollutant emissions with air fuel ratio*

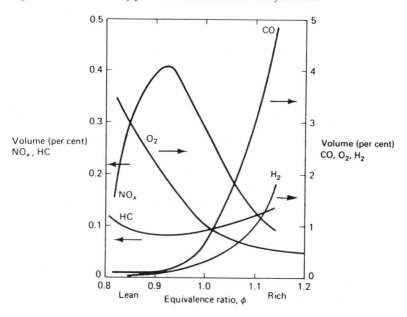

Figure 9.6   *Sources of emissions in spark ignition engines*

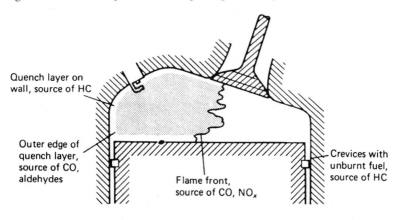

Quench layer on
wall, source of HC

Outer edge of
quench layer,
source of CO,
aldehydes

Crevices with
unburnt fuel,
source of HC

Flame front,
source of CO, NO$_x$

a spark ignition engine. By considering Figures 9.5 and 9.6, it is possible to draw some conclusions which will suggest areas that are
worthy of investigation in the future in order to reduce the level of
exhaust emissions. The output of HC and CO can be greatly reduced
by operating with lean mixtures. Alternatively, the use of exhaust gas
treatment in the form of catalytic converters can significantly reduce
the emission of HC and CO. There are a number of measures that can
be taken to reduce the emission of NO$_x$:

- flame temperature reduction;
- reduce the duration of the combustion process.

NO$_x$ production is dependent on both the rate of the reaction and the
temperature at which it occurs. This can be controlled by:

- retarding the ignition;
- exhaust gas recirculation (EGR).

Retarding the ignition process is effective as it reduces the peak pressure and temperature during the cycle. Exhaust gas recirculation
increases the concentration of residuals in the cylinder, this reduces
the temperature and flame speed. Recirculating between 5% and 10%
of the exhaust gases reduces the emission of NO$_x$ by 50%. All the measures to reduce the levels of NO$_x$ in the emissions will have a detrimental effect on the power and the economy. The use of EGR will

also have the effect of reducing the lean combustion limits, which could have an implication in the smooth running of an engine. The fact that every measure to reduce emissions has a detrimental effect on the power and efficiency may lead one to ask the question 'Do I really want to be that environmentally friendly?' When all is said and done will the average motorist be happy to pay the additional tariff for the environment? It is probably the case that the motorist will be forced to pay the price, however, the reduction in performance that may occur due to emission reducing measures can have an adverse affect on the emissions themselves.

Most of the remarks above apply to spark ignition engines. In general the diesel engine does not suffer from gaseous emissions to the same extent as its spark ignition counterpart. The most serious problem with the compression ignition engine is that of particulate emission and noise. The noise is due to combustion noise. The emission of CO is not a problem in the compression ignition engine as the engine always operates with an excess of air in the cylinder. The spark ignition engine operates much closer to the stoichiometric air–fuel ratio. The HC concentration in the exhaust gas tends to be less than the spark ignition engine, however under heavy loads the HC concentrations may approach that of a SI engine. In the case of NO_x in the exhaust gas, this is of the order of 50% of the SI.

The most easily recognized pollutant from a compression ignition engine is that of smoke. This dirty image has been one of the major drawbacks to the diesel engine. To a certain extent the situation is now being rectified, garage forecourts provide disposable gloves for customers, engines are more refined and the modern diesel is now becoming popular due to it good performance, cleaner image and fuel economy. In some countries, e.g. France, the motorist is given incentives which encourages the use of the diesel engine. These incentives are in the form of reduced fuel costs. This, together with the good fuel economy, has resulted in a major increase in the use of diesel-powered vehicles of all sizes.

The smoke originates from carbon particles formed by cracking (splitting) of large hydrocarbon molecules in the fuel rich side of the reaction zone, see Figure 9.7. The carbon particles grow by agglomeration until they reach the fuel lean zone, where they can be oxidized. The final rate of soot release depends on the difference between the rate of formation of particles and the rate of oxidation. The maximum fuel and hence maximum power is limited such that exhaust smoke is just visible. Smoke emissions can be reduced by advancing the injec-

Figure 9.7 *Sources of emissions in compression ignition engines*

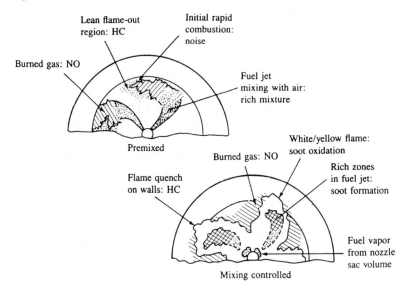

tion process or by injecting a finer spray, the latter measure can be achieved by using higher injector pressures and finer nozzles. Figure 9.7 summarizes the sources of pollutants in a compression ignition engine, the situation is not so different to that observed in the spark ignition engine.

In addition to the pollutants discussed above, carbon dioxide (CO_2) should really be added to the list. With all hydrocarbon based fuels significant quantities of CO_2 will be produced during the combustion process. Although increased levels of CO_2 indicate good combustion and will also result in reduced emission of HC, CO_2 has been identified as potentially detrimental to the environment as it is a greenhouse gas. To reduce the emissions of CO_2, more use will need to be made of bio-based fuels. More efficient operation of engines will reduce the amount of fuel used and thus reduce the emissions of CO_2.

III TOWARDS AN ENVIRONMENT RESEARCH AGENDA

Having identified the sources of emissions from engines both stationary and for transport applications, it is now possible to consider what can be done to improve the situation. There are two approaches that can be adopted, first we can look at the current technology and consider what can be done to reduce the emissions. These measures might take the form of retro-fitted devices. The second area, and probably the most important, is that of new technology.

If we consider the diesel engine, as this is now becoming more popular due to its good performance and excellent fuel economy, it is possible to see examples of both measures suggested above. The indirect injection diesel (IDI) engine has been a popular power unit for small high-speed diesel applications, typically in passenger vehicles. The direct injection (DI) engine is now becoming the more popular design due to the improved efficiency and performance. It is anticipated that the DI will eventually replace the IDI, however, this is not likely until the early part of the next century. In the meantime there are still emissions standards which have to be satisfied now. Thus research into the development of the IDI will continue for some time yet.

Taking the example of the diesel engine further, it is possible to look at the areas which are and will remain important in the development of the engine in the quest for low emissions. The main pollutants of concern from the diesel engine are:

- NO_x
- particulates

Figures 9.8 and 9.9 show the main engine parameters, which have an affect on these pollutants. First, take the case of the NO_x emissions, Figure 9.8. It is clear that this is influenced by a number of factors, all of which influence the flame temperature and reaction rates. It can be concluded that these are controlled by the conditions which dictate the mixing rates of fuel and air. Taking a similar view of the particulates, Figure 9.9, shows that again mixture formation has a significant affect. In addition to the mixing problem, the properties of the fuel and even those of the lubricating oils used need to be considered. From Figures 9.8 and 9.9 it is now possible to draw up a shopping list of research areas to be considered now and in the future.

IV RESEARCH TOPICS

Figures 9.8 and 9.9 gives details of the particular aspects of the engine geometry which influence the two major sources of pollution in diesel engines, NO_x and particulates. There are two routes to improve emissions:

Figure 9.8   *Effect of engine parameters on NO$_x$ production*

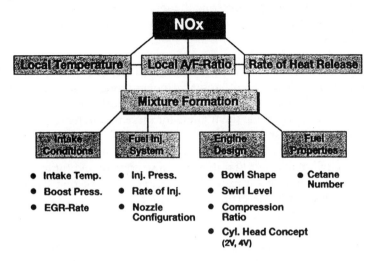

Figure 9.9   *Effect of engine parameters on particulates*

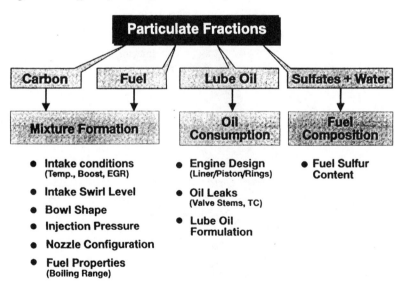

- reducing the emissions at source within the engine cylinder—this can be achieved by improving the internal processes;
- the alternative to the above is to reduce emissions by exhaust gas treatment.

If the source of the emissions is considered, i.e. in-cylinder, we find that there is a great deal of research being funded both within the motor industry and in the university system. Both industry and the research councils provide the funding for these research programmes. As stated previously, development work is being carried out at a number of levels. First, there are developments and modifications to current engines in order that they will meet current and future short-term legislation. A good example of this is the work that has been carried out with IDI diesel engine. In the IDI diesel engine air is compressed into a small pre-chamber (see Figure 9.10). The geometry of the chamber generates a strong swirling flow field. Fuel is injected into this pre-chamber where the combustion process commences. The geometry of the pre-chamber is of critical importance in controlling

Figure 9.10   *Indirect injection diesel engine geometry*

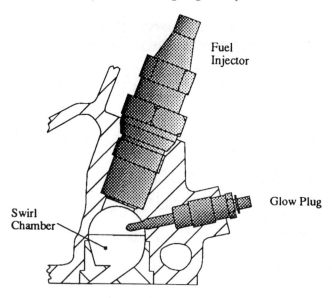

**Ricardo Comet Vb Combustion Chamber
with Crossflow Glow Plug**

the combustion process and thus emissions. Small modifications to the swirl chamber geometry (see Figure 9.10) has resulted in significant reductions in pollutant emissions. Although the DI will eventually replace the IDI, work still forms an important part of the push to reduce emissions.

In addition to this, there are new concepts in engines including a considerable interest in the potential of the two-stroke. A two-stroke engine was designed and built at the University of Bath. The engine has two cylinders and a capacity of 900 cc. It has been successfully tested and preliminary results are very encouraging. In addition, there is the orbital engine, which has been tested extensively in Australia. The advantages of the two-stroke are its high power to weight ratio and potentially low emissions, as with EGR the scavenging process will leave some residual gases in the cylinder which can significantly reduce the emission of NO_x. Most of the major manufacturers have shown an interest in the two-stroke. Many appear to be playing a waiting game, watching the others to see what will happen.

It is possible that there will be many new developments. That will reduce emissions. The introduction of electronic fuel injection and engine management systems mean that it is possible to tune engines to give low emissions while still maintaining good performance.

Inlet geometry has a significant effect on the performance. This is due to the flow structure which is set up in the cylinder. High levels of turbulence and swirling flows govern the mixing of fuel and air, and this controls the combustion process and hence emissions. In addition to the inlet geometry, there is the general cylinder geometry; the shape of the cylinder head, the position of spark plugs or fuel injectors and the piston geometry will influence the pollutants.

One of the major advances in the development of the internal combustion engine is the use of computers. Computational fluid dynamic (CFD) codes and finite element computer codes have become research tools which are now extensively used. The use of computers allows the effect of changes to engine geometry to be considered without the need to build a full-scale model the test can be performed on the computer and analysed without incurring a waste of manpower or financial resources. Figure 9.11 shows a computational fluid dynamics model of the inlet geometry of a DI engine. Use of CFD allows the flow fields to be predicted within the inlet ducts and the cylinder. A complete understanding of the flow field in the early phases of an engine design will allow an optimized geometry to be achieved as quickly as possible. Recent work is moving towards the integration of

Figure 9.11   *Two-stroke diesel engine*

predictive CFD tools into computer-aided engineering software in order to streamline the design process and reduce the time for an engine to move from concept to production. This will remove the need for costly experiments in the initial stages of the design.

With the development of sophisticated combustion models it is now possible to achieve reasonable predictions of the chemical composition of the gaseous emissions. However, there is a great deal of scope for improvements to be made which will further enhance the capabilities of such codes.

In recent years much work has been carried out on fuel quality. The emphasis has been on identifying key fuel properties which influence emissions. Thus fuels can assist in improving air quality. It is clear that

Figure 9.12 *Computational fluid dynamic model of the inlet geometry of a direct injection diesel engine*

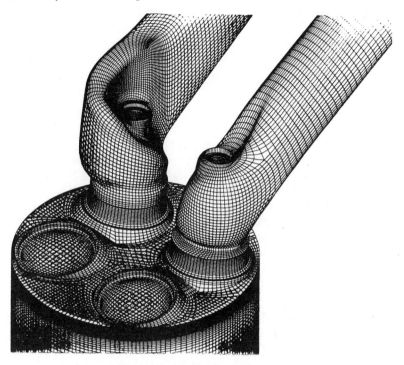

modern well-maintained diesels are low gaseous emitters, with values of HC, NO_x and CO at least as low as the competition. Most spark ignition engines require three-way catalytic converters to match the performance of a diesel engine of similar output irrespective of the fuel used.

In addition to developing conventional fuels by the use of additives, there is a growing interest in the use of vegetable oils. These are now seen as genuine alternatives to conventional fuels. The use of vegetable oil is particularly attractive to countries which do not have major resources of crude oil. It has been suggested that the use of vegetable oil can offer the following advantages:

- it is a renewable energy resource;
- high level of bio-degradability;
- little or no NO_x emissions;

- safe storage;
- no need for chemical additives;
- no net increase in CO_2.

It is possible to operate diesel engines with vegetable oils with little or no modification. Rudolf Diesel predicted more than 100 years ago that one day vegetable oil would be as important a fuel as diesel fuel itself.

V CONCLUSIONS

Above is an outline of the research agenda, perhaps a question that we should be asking ourselves is who should be involved in this process. Obviously everybody should be concerned about the environment, however, perhaps not all can contribute at the intellectual level. The impetus for the research will come from everywhere; there are many that can contribute to the development of the internal combustion engine, perhaps even looking for its replacement. The use of electric vehicles could become more widespread.

It is easy to see that engineers will be on the front line in the war against pollution, but they should not be there alone. The tools which are now available are sophisticated but that is not to say that they cannot be improved. The approach to controlling emissions should truly be a multi-disciplinary one, involving engineers, pure scientists and mathematicians. The recent initiatives which have been launched through EPSRC, the Innovative Manufacturing Initiative have highlighted the potential of this approach. With everybody working together it should be possible to solve the present-day and future problems, thus ensuring a bright and sunny future for our children.

Further Reading

B. E. MILTON (1995) *Thermodynamics, Combustion and Engines*, 1st edn, London: Chapman & Hall.
S. R. TURNS (1995) *An Introduction to Combustion*, 1st edn, NY & London: McGraw-Hill.

References

J. B. HEYWOOD (1988) *Internal Combustion Engine Fundamentals*, 2nd edn, NY & London: McGraw-Hill.

H. LIST (1993) 'Despite diesel challenges prospects are bright', *Diesel and Gas Turbine Publications*, May, pp. 26–32.

B. J. SOMERVILLE, S. J. CHARLTON, S. A. MACGREGOR and B. NASSERI (1993) 'A study of air motion in an IDI passenger car diesel engine', IMechE Conference: Experimental and Predictive Methods in Engine Research and Development, Nov.

R. STONE (1992) *Introduction to Internal Combustion Engines*, 2nd edn, Basingstoke: Macmillan.

Part III

Natural Resources and Their Management

10 Challenging Environmental Uncertainty: Dynamic Boundaries Beyond the Selfish Gene

Alan D.M. Rayner

Summary

Environmental change arises as the consequence of potentially complex interactions between living systems and their surroundings and neighbours. Foreseeing and responding to environmental change therefore depends on developing an appropriate conceptual framework for understanding these interactions and the dynamic contexts within which they occur. Current, strongly analytical models of evolutionary processes, which emphasize individual competitiveness and random chance, are based on an erroneous assumption of discreteness, and though they may provide accurate predictions in some circumstances are likely to be restrictive and distorting in others. It is argued that a more systemic paradigm is needed, based on the way energy is assimilated, conserved, distributed and recycled within dynamic boundaries whose deformability, receptivity and integrity are altered according to the availability of external free energy supplies. Plentiful supplies promote 'self-differentiation' of dissipative, relatively autonomous, competitive systems. Deprivation promotes 'self-integration', the fusion, sealing and redistribution of boundaries, allowing energy to be conserved and networked in ways that can allow large amplifications of scale. Fungi, and the way in which they are thought to respond to oxidative stress, are used as a specific example to illustrate the potential utility of this systemic approach.

I   INTRODUCTION

The interactive processes by which living systems of all kinds respond to and bring about environmental change are key to the future quality of planetary life. These processes operate at all scales from microscopic to global and at least in the short term will continue both to encompass and be greatly influenced by the thoughts and actions of human beings.

The ways in which we perceive these processes, both in ourselves and other living systems, are therefore all-important to foreseeing the potential consequences of our future action or inaction in response to environmental and social challenges. Currently, such challenges arise especially from the combined effects of population growth, changes in industrial, building and farming practices, and radical shifts in the nature, speed and accessibility of our communications infrastructures. To comprehend the full scope and ramifications of these challenges, an enormous effort will be needed to make clear the relevance of developments in particular scientific and philosophical fields to others and so provide the basis for interdisciplinary synthesis. This in turn requires individual specialists to lay bare their ideas and knowledge in such ways as to be accessible to and invite *sympathetic* comparison by others, cutting across barriers of language, ideology and technical know-how—no easy task! Here I will try to describe the systemic perspective, drawn primarily from studies of fungi, that I have developed as a biologist interested in how life forms in general produce diversity by varying the ways in which they accumulate and distribute supplies of free energy. A recurrent theme in my discussion will concern the extent to which life forms are kept competitively separate or brought co-operatively together depending on circumstances.

II   DISCRETENESS AND CONTINUITY IN LIVING SYSTEMS

During recent decades, there has been an increasing tendency to interpret how living systems interact with their surroundings and neighbours from a purely *analytic* standpoint, i.e. in terms of the generation and selection of discrete, free-ranging, self-centred survival units in the form of genes and individuals (Dawkins, 1976). The dynamic processes underlying change are thereby treated as though *administered*, at least to a good approximation, solely by *external* forces acting on spatially and temporally finite *objects* with absolute beginnings and endings, i.e. *absolute boundaries*. The path of change can thence be calculated, at least in the short term, with a precision and certainty that is often celebrated as a source of great scientific strength. The evolutionary 'River out of Eden' is perceived as 'just bytes and bytes of digital information' (Dawkins, 1995, p. 19). Success in life is widely viewed either as the consequence of individual competitiveness or of random chance in a lottery of fortune and misfortune.

From a *systemic* perspective, however, this discretist approach is potentially very restrictive and misleading because all known life forms depend on *internal* energy conversions if they are to prosper and change with time. As *dynamic systems*, life forms *assimilate* sources of free energy from their surroundings and *distribute* this energy into proliferation and movement. They achieve this by possessing *boundaries* that *to some degree*, and ranging from subcellular to ecosystem in scale, *both* resist *and* permit energy exchange with their surroundings and neighbours (Rayner, 1997a). These boundaries cannot be absolutely sealed if they are to permit assimilation, and so cannot entirely prevent leakage or incursion. By the same token, they cannot be absolutely absent or lost if they are to prevent irreversible discharge of a living system's contents: continuity of boundaries is as essential to the continuation of life as the inheritance of genes.

All life forms are therefore necessarily leaky containers that are thereby *to some degree* both separate from and in communication with their progenitors, offspring and neighbours, as well as their non-living surroundings. Moreover, since the external supply of free energy varies, both in space and time, they have to be capable of minimizing losses in energetically restrictive environments through various means of containment, as well as maximizing gains in unrestrictive environments by increasing their assimilative surface. This requires changes in the configuration and properties of living system boundaries according to circumstances. These boundaries cannot therefore be absolutely fixed any more than they can be absolutely sealed: rather they are dynamic, reactive interfaces, forever in some degree of flux.

III   INDIVIDUAL TRAJECTORIES: EXPLORING DYNAMIC CONTEXTS

To realize the full scope afforded by a systemic perspective, we therefore need to shift away from viewing what we call 'individuals' as *fully* determinate entities within absolute boundaries of space and time. Rather we should view individuals as indeterminate, interactive *trajectories* with varied degrees of freedom within dynamic boundaries.

One way of appreciating what this shift entails is to ask yourself a variety of simple questions. Where have I been? How did I get there? What conditions did I encounter? How was I resourced? Whom did I interact with? How did I manage to thrive *and* survive? How much of this could have been predicted?

Answering these questions involves picturing some kind of map, i.e. a *trajectory* or 'world line', that you have moved along bodily through space and time, interspersed with varying periods of residence and migration.

This trajectory defines the *dynamic context* of your life—the shifting arena within and across which you have exchanged energy, the source of your driving force, with your surroundings. It is *indeterminate* in being capable of expanding and changing direction indefinitely—at least until death, when it becomes assimilated into other contexts. It is also to a greater or lesser extent *unpredictable*, because it can both influence and be influenced by the changes of circumstance that characterize life in the inconstant and uncertain environments of the real world. In other words, trajectories are subject to *feedback*, in much the same way as a river whose course both influences and is influenced by the landscape it flows through. Tiny changes in circumstances can be amplified into fundamental alterations in direction. Success and failure are most fundamentally about serendipity and misadventure—emerging in the right or the wrong context.

Moreover, it is the boundaries of your dynamic context, rather than your bodily disposition at a particular moment, which best define yourself as a dynamic, interactive system. These boundaries have both regional and topographical aspects. Regional boundaries are analogous to the watershed of a river basin and define overall domain or territory. Topographical boundaries are defined by the trajectories or pathways followed within this domain and are analogous to river tributaries and distributaries. (Please note that in this systemic sense, 'context' is an interactive spatiotemporal domain or 'field', closely allied to the concept of 'niche', that is *incorporated* by a living system; it does *not* equate with *external* environment, separate from the living system.)

Here it is important to acknowledge just how difficult the technical and imaginative challenges are to seeing animals like ourselves as trajectories in a dynamic context. Not only is it generally necessary to develop some kind of moving picture imagery of an animal's life's journey but it is also vital to be able to merge successive time frames to reveal the trajectory that maps this journey, and its intersections with others.

There are, however, some life forms which make the systemic, contextual view much more accessible. This is because these life forms *grow* rather than move bodily from place to place. Consequently, their contextual boundaries coincide directly with their body boundaries and so can be immediately visualized.

We are utterly surrounded by and dependent upon these life forms, which include many plants and fungi. However, perhaps because they seem a bit alien to us animals, we have remained largely unreceptive to the insights into indeterminate organization within dynamic contextual boundaries these forms provide for us.

Among the most striking demonstrations of the role of indeterminate organization within inconstant and therefore unpredictable environments are provided by fungi. Although many people tend to equate fruit bodies—toadstools, brackets, puffballs and the like—with the whole of a fungus, these structures are, like streetlamps in a city, actually only the outward signs of underlying dynamic networks. Composed of microscopic, interconnected, branching, protoplasm-filled tubes (hyphae) that extend at their tips, it is these dynamic networks, known as mycelia, which locate, take in and distribute the often patchy supplies of nutrients, oxygen and water that furnish the growth needs of a fungus. In so doing they provide a hidden but vital infrastructure that interconnects the lives and deaths of other organisms within our natural ecosystems in innumerable and often surprising ways.

Purely by responding to local circumstances, and without any administrative centre, fungal mycelia are capable of generating sustainable systems that are reinforced along paths of successful exploration. As illustrated in Figure 10.1, their organization provides them with the versatility to interconnect distinctive local sites in heterogeneous environments by shifting between energy-gathering, exploring, conserving and recycling states according to the external availability of resources. Moreover, they can do this over spatial scales ranging from μm to perhaps even km, and temporal scales ranging from seconds to millennia. The largest confirmed mycelial network occupies 15 hectares and is estimated to be aged 1,500 years (Smith *et al.*, 1992).

However, as already implied, it is vital to recognize that this kind of organization, while superlatively exemplified by fungi, is by no

Figure 10.1   *Network produced by the mycelium of the magpie fungus* (Coprinus picaceus) *grown in a matrix of interconnected 2 × 2 cm chambers alternately containing high and low nutrient media, starting from high nutrient medium in the central chamber. (Photograph by Louise Owen and Erica Bower. See Rayner, 1996b for further information.)*

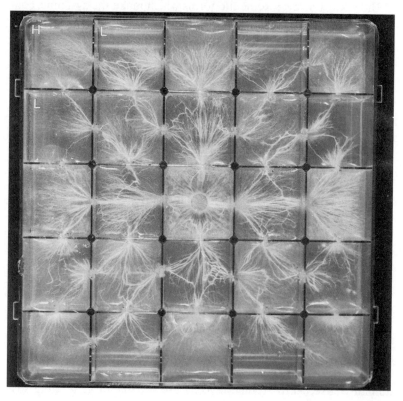

means unique to these organisms. Closely similar patterns are produced by the roots and stolons of many clonal plants, like ground ivy, couch grass and strawberries. The same can be said of nervous systems, blood systems and social gatherings of animals (including people), where individual trajectories diverge, converge, follow and reinforce one another along paths of least resistance (Figure 10.2; Helbing *et al.*, 1997).

Neither is this kind of organization relevant only to the disposition of life forms in the short term. It is deeply embedded in all manner of historical structures, including phylogenetic trees and genealogies, which reflect the ability of past contexts both to create and constrain

Figure 10.2   *The 'great trek'—a herd of wildebeest on the Serengeti plain in E. Africa migrates along well-worn trails towards river lands as the dry season advances. (Based on an aerial photograph, from Rayner, 1997a.)*

future contexts in fundamentally unpredictable—though not incomprehensible—ways. It is also relevant to the understanding of many kinds of differentiation and integration processes involved in thought, communication and economics, and the patterns generated by many non-living systems.

In short, the patterns this organization gives rise to are a universal feature of indeterminate—dynamically bounded—systems. How, then, can their physical origin best be understood, and what bearing does this origin have on the challenge of environmental uncertainty?

IV   CONTEXTUAL DYNAMICS: BOUNDARIES AND THE SOUCE OF PHENOTYPIC VARIETY

The source of the immense variety of physical forms and behaviours (i.e. 'phenotypes') produced by living systems, though critical to understanding the causes and consequences of environmental change, is perhaps the most unresolved issue in evolutionary biology. Currently, the predominant approach to this issue continues to be analytical, based on the paradigm of 'genetic determinism'.

Genetic determinism assumes that phenotype is *hierarchically generated* from genetic information content (i.e. genotype), subject to *moderation* by the *external* environment. This follows from presupposing, as an erroneous product of discretism, that only genes, and not context, are transferred between generations. Also, in itself, it says nothing about *how* phenotypes actually arise from gene products, nor about how the environment can exert its moderating influence.

From a systemic viewpoint, however, it is inescapable that phenotypes arise from the way that the physical properties of the dynamic contexts and boundaries of living systems affect patterns of energy uptake and transfer. Many of these properties derive from materials or energy sources—e.g. water, air, minerals and light—that are not encoded in genes, but are nevertheless salient in *moulding* the dynamic interplay between genetic information and context into varied forms and behaviours.

In more ways than one, the origins of phenotypic diversity can therefore be traced to the interaction between two *interdependent* parallel strands—a contextual strand of interconnected, dynamic, living system boundaries, and a genetic strand of variations in DNA sequence. The continual emergence and reconfiguration of contextual boundaries provides the evolutionary *opportunities* (dynamic niches) for genes to proliferate in an indefinite, autocatalytic feedback process: the emergence of a tree produces a context for a climbing plant to evolve in.

So, the nature of the informational traffic both influences and is influenced by the nature of the contextual highway—just as the form of a river depends on the interaction between its ingredients and its banks. Driven primarily by energy input, *both* nature (internal influences) *and* nurture (external influences) combine and inextricably intertwine at the dynamic contextual boundaries that face *both* the inside *and* the outside of living systems.

Phenotypic diversity and responsiveness may therefore best be understood in terms of the systemic paradigm of 'contextual dynamicism'. Here, phenotype is envisaged to be a contextual function of variations in gene content, gene expression and external environment (Davidson *et al.*, 1996; Rayner, 1996a, 1997a). Genes can thereby be understood as an inheritable—though *not*, in isolation, sustainable—means by which particular sets of contextual properties are specified, *reiterated* and diversified. However, it is the dynamic boundaries and the materials and energy sources they contain and interact with, that enable and define the evolutionary action. But, the question remains —*how*?

V BOUNDARIES OF ORDER, ORGANIZATION AND CHAOS

Recent decades have witnessed significant developments in the way that pattern-generating processes in dynamic physical systems can be understood and modelled mathematically. These developments are encompassed within an array of interrelated concepts, variously described as nonlinearity, chaos, complexity, fractal geometry and self-organization. Their relevance to the systemic origins of phenotypic variety is potentially profound, but still seems not to have been widely acknowledged and incorporated into biological philosophy and methodology.

Self-differentiation: the Route to Incoherence

Many of the developments just referred to arise from the effects of two kinds of feedback and their counteraction. Positive feedback, autocatalysis, arises from the ability of a system to amplify itself using energy input from its local environment. This ability generates an expansive drive which, if unconstrained, causes the system to increase exponentially. Negative feedback damps down expansive drive by directly or indirectly increasing resistance or dissipation as input increases.

The counteraction between positive and negative feedback causes systems to be nonlinear (non-additive) and to become unstable if the rate of input exceeds a critical threshold. Below this threshold, the counteraction causes a smooth build up to a dynamic equilibrium at which there is no net increase in the system's expansion: the system then remains, in effect, self-contained. Above this threshold, the system becomes 'forced' and hence prone to subdivide, by means of a series of bifurcations, into increasing numbers of recurrent subdomains or states. Above a yet higher threshold, the subdivisions can cease to occur recurrently. Instead, the system traverses what approaches an infinite variety of states—in a manner which is apparently erratic and extremely sensitive to initial conditions, and therefore unpredictable in the long term. This is deterministic chaos.

An implicit—but all too commonly not an explicit—feature of physical systems that exhibit non-linear dynamics is the presence of one or more dynamic boundaries. The very term, 'feedback' is defined most simply as an influence of the output from on the input to a system and therefore implies a reactive interface that mediates the extent of this influence. Without a boundary—whether due to attraction or constraint—that allows assimilation but prevents instantaneous dispersion, there can be no autocatalysis and no containment.

The fact that the important role of dynamic boundaries is often overlooked, has led consciously or unconsciously to discretist interpretations of non-linear systems and thence to some possibly profound misconceptions. These misconceptions arise because attention is focused on the behaviour of individual components of the systems rather than the boundaries which shape and are shaped by these behaviours.

Perhaps foremost among these misconceptions occurs in what has been termed 'self-organization theory' and its attendant metaphor of 'order out of chaos' (Prigogine and Stengers, 1984). For self-organization to occur, it has generally been considered necessary for the systems to be thermodynamically open and far from equilibrium, so that they can be sustained by high rates of energy input. Under these circumstances they produce emergent structures or patterns that maximize the conversion of free energy input to entropy and so are described as 'dissipative' (Prigogine and Stengers, 1984). Moreover, since emergence occurs in what appears to be a previously patternless or structureless domain, it is assumed to originate from chaos or even randomness. Examples commonly used to illustrate this idea include 'random' mixtures of autocatalytic ('activator') and constraining ('inhibitor') chemicals, and 'random' arrays of social organisms (e.g. slime mould amoebae, ants). These systems generate annular and spiral patterns if suitably prompted by local perturbations (e.g. Goodwin, 1994). They are commonly modelled mathematically using reaction-diffusion equations.

From a systemic perspective, randomness is not, however, the same as chaos and, at any particular scale of reference, it is the boundary of a system which represents its relative order and dynamic state. The effect of introducing free energy into a dynamically bounded system is, directly or indirectly, to cause an expansion of the system's boundary. If the rate of input is below a threshold value, the boundary expands smoothly, retaining its symmetry and minimizing its surface area and consequent dissipation to its surroundings. However, if the rate of input exceeds the 'throughput capacity' defined by the resistances imposed by the system's boundary, the system begins to lose coherence by breaking symmetry and generating emergent structure. It first polarizes and then subdivides to produce more and more dissipative (and assimilative) free surface (Figure 10.3).

All this emergent, increasingly complex structure, the most extreme form of which is chaotically distributed, represents proliferated boundary—and hence, according to present definitions, *increased* order. However, the origin of this order is not disorder, but a highly integrated, coherent initial state that is capable of assimilating free energy at sufficient rates to break its own symmetry. I suggest this initial state may well be described as more highly *organized*, notwithstanding previous indiscriminate use of that term.

The dissipative structures of self-organization theory therefore emerge not from randomness but rather from an initially coherent, bounded state that enables the input of energy and associated autocatalytic processes to be initiated. The 'seeding' of this state into an energy-rich field—the latter itself necessarily being bounded or extending to infinity—is the perturbation required to set the self-organizing processes into motion.

Like the packaging that is used to distribute all kinds of commodities, the order invested in boundaries is energetically costly. There are two reasons for this. The first reason is the high rate of free energy input required to cause systems to become unstable and break symmetry. The second reason is the increased dissipative free surface presented by proliferating boundaries. This increased surface both renders the system more susceptible to random environmental influences and counteracts the destabilizing effects of the input of free energy which causes the boundaries to proliferate in the first place. Correspondingly, more erratic but less labile structures emerge—a fact which is overlooked by many purely deterministic mathematical models.

Given that boundary-proliferation can only be sustained by continuing energy input, an important question is what happens to dissipative structures when external supplies of free energy are restricted? A related, fundamentally important, question is what is the origin of the initial coherent state from which dissipative structures emerge in energetically unrestricted environments?

Essentially, if external energy supplies are withdrawn from a

Figure 10.3 *The role of dynamic boundaries in the production of dissipative structure ('order') and coherent 'organization'. Assimilation of free energy into a coherent initial state (C) results in the proliferation and subdivision of boundary (dissipative structure) by 'self-differentiation' (D). Irreversible decay or degeneration (DE) of this structure in the absence of energy replenishment leads to random disorder (R). 'Self-integration' of this structure by boundary-fusion, boundary-sealing and boundary-redistribution minimizes its dissipative free surface, enabling it to reconfigure into coherent initial states or persistent networks (N). (From Rayner, 1997b.)*

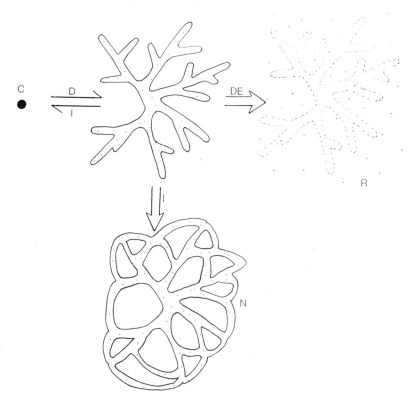

dissipatively structured system, the long-term survival of the system (or part of it) rests on a stark choice (Figure 10.3). First, the system may continue to dissipate. Alternatively, it undergoes processes that minimize exposure of free surface. The first option leads to dissolution, an irreversible decay into an entropic state. The second option results in a *reduction* of *order* commensurate with an *increase* in *organization* (coherence). The sustainability and persistence of life forms in energetically inconstant environments depends on this second option, which involves three dissipation-minimizing processes: boundary-fusion, boundary-sealing and boundary-redistribution. I term this second option 'self-integration', to distinguish it from the emergence of dissipative structures, which I term 'self-differentiation'. 'Self-differentiation' then implies the division of self boundaries into progressively less coherent, potentially competitive, energetically extravagant sub-units. 'Self-integration' implies the unification of self boundaries into a more coherent, less dissipative organization. Self-differentiating processes have dominated evolutionary, environmental and socio-economic thinking for much of the twentieth century. By contrast, self-integration has largely been ignored or taken for granted. This neglect goes hand-in-hand with discretism and a consequent over-emphasis on competition as the prime driving force for adaptive refinement in a fuel-rich evolutionary economy.

Self-integration: Retaining and Regaining Coherence

Since the maintenance of boundaries, as dissipative free surface, is energetically costly, any processes that minimize this surface are energy-saving—and may even be energy-yielding. Figure 10.4 shows how the three such self-integrational processes identified above operate over the course of indeterminate individual life spans and life cycles to enable living systems to gather, conserve, explore for and recycle sources of free energy according to local circumstances in inconstant environments. Please compare this figure with the biological examples shown in Figures 10.1 and 10.2 and the depiction of order, organization and chaos in Figure 10.3.

Boundary-fusion both reduces dissipative free surface and releases energy that was previously contained in this surface. It can convert a dendritic branching system with resistances to throughput in series, to a network with resistances at least partially in parallel. It thereby increases the throughput capacity and restores the symmetry of the system, making it more retentive and less prone to proliferate branches. At the same time it enables the system to amplify its organizational scale, through enhanced delivery to sites of emergence of distributive or reproductive structures on its boundary. Successive rounds of boundary-proliferation and boundary-fusion may also have provided scope for evolutionary amplifications of scale from microbial biofilm systems around $100\,\mu m$ high to redwood and eucalypt forest systems about $100\,m$ high.

Figure 10.4   *The interplay between self-integration and self-differentiation to produce distinctive organizational states in resource-rich (stippled) and resource-restricted environments. The interplay enables energy to be assimilated (allowing regeneration of boundaries), conserved (by conversion of boundaries into impermeable form), explored for (through internal distribution of energy sources) or recycled (via redistribution of boundaries) according to circumstances. Thin lines indicate permeable contextual boundaries, thick lines impermeable boundaries and dotted lines degenerating boundaries. (From Rayner, 1997a.)*

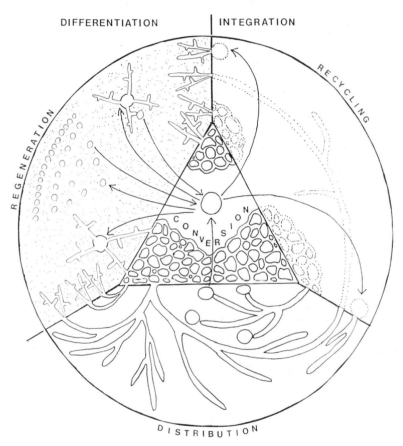

Boundary-sealing involves various ways of impermeabilizing or 'insulating' living system envelopes. Sealing a fixed boundary results in the production of dormant survival structures. Sealing a deformable boundary results in the emergence of distributive structures that either serve reproductive or explorative/migratory functions. Since the sites of input to these structures are distal to their sites of proliferation, their branching pattern will be distributary-like, contrasting with the tributary-like branching pattern of assimilative structures.

Boundary-redistribution involves the transfer of resources from degenerative to generative sites. Although degenerative processes are often viewed as self-destructive and wasteful, they may actually have the opposite effect. This is because they break the 'gridlock' that can develop in an over-retentive, networked system, and so allow proliferation to continue, a feature which has been demonstrated mathematically by Davidson *et al.* (1996, 1997). As will next be described, they can also allow systems to isolate themselves from potential takeovers.

VI BOUNDARIES IN COMPETITION, CONFLICT AND CO-OPERATION

Fundamentally, then, the ability of individual living systems to generate phenotypic variety depends on the relative *deformability*, *receptivity* and *integrity* of the contextual boundaries that both resist and enable energy input and dissipation. This is true of all life forms.

These contextual boundary properties are also of fundamental relevance to the potentiality for competition, conflict and co-operation between neighbouring organizations. Some of the general principles underlying this statement are shown in Figure 10.5, and an illustrative fungal example is shown in Figure 10.6.

When the boundaries of two life forms occupy the same context, they will, at least initially, *compete* for resources. If they come into actual physical contact, they may even compete for resources contained within one another—as when a seagull steals bread out of the mouth of another, or cancer cells proliferate in a human body. The competition may culminate in monopoly, stalemate, mutual destruction or subdivision into specializations.

Such competition can result in 'refinement', a honing of boundaries that maximizes the chances of success. However, it is also uses up irrecoverable amounts of energy, and whilst it may originate in 'choice' it commonly ends either in no choice or an array of incomplete 'sub-choices'—as any bewildered supermarket shopper knows. When intense it also leaves little room for manoeuvre—experimen-

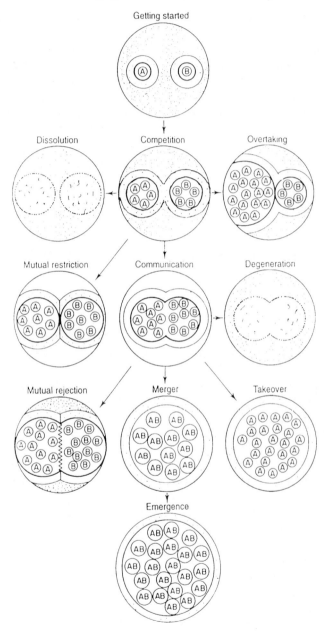

tation with boundary conditions—because any significant departure from existing practice, any 'error', is unlikely to succeed in the short term.

However, this competitive phase can be superseded when life forms join forces by integrating their boundaries, so producing partnerships and networks in which there can be complementation and pooling of resources. In those cases where a smaller entity becomes enclosed within a larger one, much depends on what then happens to the smaller entity's boundary. If it breaks down, the smaller entity is consumed. If it remains intact, the smaller entity becomes an inhabitant of the larger entity, now its host, in which it may live either as a parasite or as a beneficial partner. On the basis of such partnerships have been built coral reefs, forests and perhaps even all creatures, great and small—animals, plants and fungi—that are made up of what are known as 'eukaryotic' cells.

Complete integration does not, however, solve the problem of excessive competition. Networks can become 'establishments' that

◄─────────────────────────────────────

Figure 10.5 *Corporate interplays. Two systems, containing components (A and B) with similar but not identical capabilities and requirements for resources begin to establish themselves in the same arena, represented here as a circle with densely stippled contents. Each system is surrounded by a region of 'influence' (lightly stippled) which it draws on to support its activities. If the net income from this region balances net expenditure, the system will maintain but fail to expand its boundaries. If there is a 'profit margin' (income exceeds expenditure), the system will grow. However, if there is a deficit, the system will lose viability.*

Both systems grow until their regions of influence overlap and begin also to become limited by the boundaries of the arena. The resulting competition can reduce profit margins to the point where no further expansion is sustainable (leading to mutual restriction), or beyond this point and into deficit (leading to dissolution). However, if one system proves to be more competitive, it will begin to monopolize the arena as it erodes the other's region of influence (overtaking).

On the other hand, if the systems integrate their boundaries and start to communicate, they will have the opportunity to operate as one and so expand their mutual influence. Much now depends on the way the components interact. If they compete or interfere, the resulting incompatibility may culminate in extensive degeneration, mutual rejection or takeover. If they complement, they may not only allow the corporate structure to fill the arena to its fullest potential but also enable it to expand the boundary of the arena itself, by means of innovative interactions.

The same principles apply to all kinds of living systems that possess dynamic boundaries, from cells to societies. (Modified from Rayner, 1996b.)

Figure 10.6   *Brute force in an interaction between mycelial systems of two different species of basidiomycete fungi when grown next to one another in a Petri dish.* Hypholoma fasciculare *(lowermost) has produced invasive mycelial cords at the 'weak' centre of the interaction interface, but in so doing left itself open to a flanking 'pincer attack' by dense invasion fronts produced by its opponent,* Coriolus versicolor. *(From Griffith et al., 1994)*

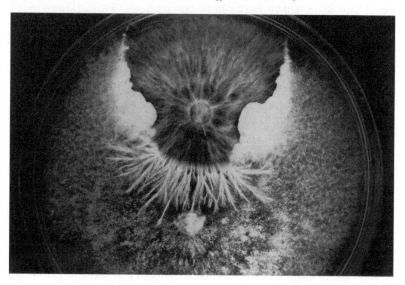

keep resources to themselves rather than allocate them to 'explorers' on their boundaries. The 'fairy rings' formed by fungi illustrate this: mycelial networks can only continue to expand outwardly through the disintegration of their interiors; sooner or later past endeavour has to be relinquished if it is not to become stifling (Davidson *et al.*, 1996, 1997). Furthermore, integration into a network opens up the possibility of competition between entities within the network. Such competition can result in social parasitism—the emergence of local 'power drains' that monopolize resources. This possibility has been clearly demonstrated in ant colonies and fungal mycelia, and is only too familiar in many human organizations. It can be prevented by a degenerative 'rejection reaction' which closes off the communication channels. True co-operation requires the assertion of 'individual dignity'—an ability to preserve as well as combine boundaries. That way, the forces of integration and differentiation can continue to negotiate a path into the uncertain future.

VII FEEDBACK MECHANISMS: THE DYNAMIC INTERPLAY OF THE ELEMENTS

Having hopefully identified the generic boundary properties governing contextual dynamics, the next step is to characterize the specific feedback mechanisms responsible for varying these properties in particular cases. These mechanisms are likely to differ greatly depending on the nature and scale of the systems concerned, but all are likely to involve an interplay between a system's contents, its surroundings and its energy source. I will use fungal mycelia as a particular example to illustrate this interplay.

The contextual boundaries of fungal mycelia are defined by the envelopes of hyphal tubes. These envelopes consist of a combination of cell wall and plasma membrane, and serve to regulate, respond to and canalize the uptake and distribution of water and nutrients through the system.

Given that the deformability, permeability and integrity of hyphal envelopes will depend primarily on their chemistry, the key questions to ask are how does this chemistry affect, and how is it influenced by the external and internal environments of hyphae?

Since the energy which drives hyphal proliferation derives most fundamentally from external sources of reducing and oxidizing power, a logical starting point for understanding this feedback would be to consider how boundary chemistry is affected by redox conditions. Not least, it may be important to be aware of the potency of what might well be described as the world's first dangerously addictive drug—oxygen!

Oxygen is a gas upon which all the most complex forms of life have come to depend, that provides a huge energy boost, but which has to be consumed in regulated doses and which is likely to kill them in the end! The reason that oxygen possesses this Shiva-like role as provider and destroyer of life lies in its affinity for electrons, which it accepts one at a time in the course of its reduction to water. In the process, reactive oxygen species (ROS) are formed which are capable of destroying the chemical order of living protoplasm, through the generation of free radicals (species with one or more unpaired electrons) (Halliwell and Gutteridge, 1989).

Here, important evidence has been presented by Hansberg and Aguirre (1990), who found that the development of a 'hyperoxidant state' is a necessary prelude to aerial mycelium and spore formation in the ascomycete fungus, *Neurospora crassa*. This is an unstable state in which the capacity of protoplasm to neutralize ROS is exceeded. Unless mitigated in some way, it is liable to lead to protoplasmic degeneration. It will be promoted by any factors that diminish availability of reducing power, enhance exposure to oxygen (especially in the gaseous phase, as in terrestrial habitats) or impede respiration. On the other hand it will be attenuated by any mechanisms that maximize resource uptake while minimizing intracellular oxygen concentrations.

From this perspective, mycelial systems have four main ways of responding to the threat and promise of oxygen, each with contrasting effects on boundary chemistry. First, they can assimilate nutrients in solution from plentiful external supplies through hydrophilic, permeable boundaries: in so doing they acquire respirable substrate and the ability to proliferate as dissipative systems—but only as long as nutrients are replenished. Secondly, they can neutralize intracellular ROS and molecular oxygen by means of antioxidant enzymes, pathways and metabolites—many of which are currently classified under the general heading of 'secondary metabolism'. Thirdly, they can produce a relatively oxygen- (and thereby also solute- and water-) impermeable boundary. This can be achieved by fusion and aggregation of hyphal envelopes (so restricting proliferation of branches) and the generation and oxidative cross-linking of hydrophobic phenolic, proteinaceous and lipid compounds in the presence of phenoloxidase and peroxidase enzymes. Fourthly, they can actively enable or passively allow a hyperoxidant state to arise and lead to protoplasmic disorder and cell degeneration (Rayner, 1996a, 1997a).

Onset of the hyperoxidant state, due to being unable, *for whatever reason*, to reduce intracellular oxygen fully to water via the respiratory chain, may therefore be an important, and possibly the most fundamental cue for initiating self-integrational responses in mycelia. Moreover, the degree of oxidative stress would determine the kind of response such that degenerative processes would be initiated above a high stress threshold, whereas protective mechanisms would come into play above a lower threshold.

This scenario, if proved, would provide a framework for explaining all the observed patterns of growth and interaction of fungal mycelia. It demonstrates how systemic considerations of boundary properties may afford new insights into the evolutionary role of non-genetic contextual components—in this case, water, oxygen and reducing power—that are all too often taken for granted or left out of the equation connecting genotype and phenotype altogether.

VII   CONCLUSION

The systemic approach of contextual dynamicism resolves many of the paradoxes that arise from trying to interpret life forms as assemblies of discrete, freely selectable units. However, it also implies *some* loss of the controllability, predictability and quantifiability that many view as the foundation of 'strong' science and its technological and environmental application. Ultimately, a choice may be necessary. Is it apt to continue to view life aridly and confrontationally from outside, as centrally controlled and predictable within fixed boundaries? Or is life really wild, wet and full of surprises as it flows within dynamic boundaries? The humility of the second view may prove to be an essential ingredient of a future in which we strive compassionately to engage with rather than gain ascendancy over one another and the rest of nature.

Further Reading

J. COHEN and I. STEWART (1994) *The Collapse of Chaos*, London: Penguin.
P. COVENEY and R. HIGHFIELD (1991) *The Arrow of Time*, London: Flamingo.
E. DE BONO (1990) *I Am Right—You Are Wrong: from This to the New Renaissance: from Rock Logic to Water Logic*, London: Viking.
G. JOHNSON (1996) *Fire in the Mind*, London: Viking.

References

F. A. DAVIDSON, B. D. SLEEMAN, A. D. M. RAYNER, J. W, CRAWFORD and K. RITZ (1996) 'Context-dependent macroscopic patterns in growing and interacting fungal networks', *Proceedings of the Royal Society of London*, series B, 263, pp. 873–80.
F. A. DAVIDSON, B. D. SLEEMAN, A. D. M. RAYNER, J. W. CRAWFORD and K. RITZ (1997) 'Travelling waves and pattern formation in a model for fungal development', *Journal of Mathematical Biology*, 35, pp. 589–608.
R. DAWKINS (1976) *The Selfish Gene*, Oxford: Oxford University Press.
R. DAWKINS (1995) *River Out of Eden*, London: Weidenfeld & Nicolson.
B. GOODWIN (1994) *How the Leopard Changed Its Spots*, London: Weidenfeld & Nicolson.
G. S. GRIFFITH, A. D. M. RAYNER and H. G. WILDMAN (1994) 'Interspecific interactions and mycelial morphogenesis of *Hypholoma fasciculare* (Agaricaceae)', *Nova Hedwigia*, 59, pp. 331–44.

236   *Challenging Environmental Uncertainty*

B. HALLIWELL and J. M. C. GUTTERIDGE (1989) *Free Radicals in Biology and Medicine*, 2nd edn., Oxford: Clarendon Press.

W. HANSBERG and J. AGUIRRE (1990) 'Hyperoxidant states cause microbial cell differentiation by cell isolation from dioxygen', *Journal of Theoretical Biology*, 142, pp. 201–21.

D. HELBING, J. KELTSCH and P. MOLNÅR (1997) 'The evolution of human trail systems', *Nature*, 388, pp. 47–50.

I. PRIGOGINE and I. STENGERS (1984) *Order out of Chaos*, London: Heinemann.

A. D. M. RAYNER (1996a) 'Antagonism and synergism in the plant surface colonization strategies of fungi'. In Morris, C. E., Nicot, P. and Nguyen, C. (eds), *Microbiology of Aerial Plant Surfaces*, pp. 139–54, London: Plenum Press.

A. D. M. RAYNER (1996b) 'Interconnectedness and individualism in fungal mycelia'. In Sutton, B. C. (ed.), *A Century of Mycology*, pp. 193–232, Cambridge: Cambridge University Press.

A. D. M. RAYNER (1997a) *Degrees of Freedom: Living in Dynamic Boundaries*, London: Imperial College Press.

A. D. M. RAYNER (1997b) 'Evolving boundaries: the systemic origin of phenotypic diversity', *Journal of Transfigural Mathematics*.

M. L. SMITH, J. N. BRUHN and J. B. ANDERSON (1992) 'The fungus *Armillaria bulbosa* is among the largest and oldest organisms', *Nature*, 356, pp. 428–31.

11 Construction and Deconstruction: Ecological Politics After the End of Nature

Ingolfur Blühdorn

Summary

In recent years environmental sociologists have become increasingly interested in the question why certain phenomena of environmental change raise massive social concern, while other arguably much more important issues remain largely unnoticed. The relationship between physical environmental change and the societal reaction it triggers seems to be largely contingent; it is definitely underresearched. In response to this observation, a number of environmental sociologists have adopted a social constructionist approach, i.e. they regard nature and ecological problems as social constructions rather than as realities external to social discourse. A social constructionist perspective can help to explain how ecological claims are made and sustained, and to what extent the shaping of environmental agendas may be understood as an immediate response to physical conditions in the environment. This chapter argues that social constructionists have developed a very promising line of argument, which should, however, be refined in order to realize its full potential.

I INTRODUCTION

Since the second half of the 1980s we can safely say that the ecological issue is firmly rooted on the political agenda and an uncontested component of mainstream politics. Most significantly in the industrialized countries, but beyond these also in the NICs and increasingly even the Third World, ecology has emerged as a new 'ideological masterframe', a 'non-controversial collective concern' (Eder, 1996). This

237

is, of course, not to say that, finally, ecological problems are fully under control and late modern societies are safely on the road towards sustainability. But at least the confrontational style of environmental politics, which was dominant in the 1970s and for the better part of the 1980s, has been replaced by a much more co-operative and consensus-oriented style, and although global warming, the loss of biodiversity, soil erosion, increasing population levels, etc. present seemingly insurmountable problems, there is significant optimism that a coalition between politics, industry and an environmentally educated public can bring about the turn towards sustainability. Taking account of this development, Klaus Eder announced the advent of 'post-environmentalism', which distinguishes itself from earlier phases of eco-politics in that it goes beyond environmental mobilization and turns 'the collective consensus on a green agenda' (Eder, 1996, p. 183) into political practice.

The increasing incorporation of environmental considerations into economic, social and political decision-making processes has been described as the *ecological modernization* of late modern society. In recent years a number of theorists have adopted this concept and tried to spell out what kind of societal transformation it implies (e.g. Huber, 1993; Jänicke, 1993; Hajer, 1995; Mol, 1996; Christoff, 1996; Spaargaren, 1996). Ecological modernization has been defined as the attempt to modernize late industrial societies 'by repairing a structural design fault of modernity: the institutionalised destruction of nature' (Mol, 1996, p. 305). Ecological modernization hopes to accomplish a new state of stability, often described as sustainability, which restores 'the balance between nature and modern society' by means 'reembedding' (ibid., p. 306) the latter into the former.

A closer look at the reality of environmental politics in the 1990s, however, immediately reveals the problems implicit in the idea of ecological modernization. Contradictory opinions not just about global phenomena like ozone depletion or climatic change, but even about local problems like waste management, energy provision or air pollution translate into multiple and often mutually exclusive views about what measures should best be taken and who should actually take them. On the one hand, contemporary eco-politics is characterized by almost ubiquitous green commitment, but, at the same time, it is paralysed by an increasing fragmentation of the ecological consensus. Whether the process of ecological modernization will really take society any closer to the utopia of ecological sustainability is, therefore, a separate issue. If this is the aim of ecological modernization, which may of course be contested (Blühdorn, 1999), its success strongly depends on the availability of a consensus about ecological goals and strategies. In late modern societies, however,—even if we call them 'post-environmentalist'—such a consensus does not seem to be easily achievable. In a recent critical study, Maarten Hajer described this dilemma as the emergence of a 'new environmental conflict', which 'no longer focuses on the question of whether there is an environmental crisis', but 'is essentially about its interpretation' (Hajer, 1995, pp. 13f).

What I want to do in this chapter is firstly to conceptualize this difficulty in terms of the *abolition of nature*. Secondly, I want to explore to what extent environmental sociologists have taken account of this abolition of nature. In particular, I want to discuss the so-called *constructionist* approaches offered by some environmental sociologists and the criticism levelled against them by their *realist* opponents. And thirdly I want to explore how the constructionist approach may be further developed in order to provide a more plausible explanation of the relationship between the natural environment and the setting of political agendas. A refined constructionist research agenda may help us to get a fuller understanding of the societal process of ecological modernization and the prospects for environmental politics beyond the millennium.

II THE ABOLITION OF NATURE

The idea of the abolition of nature is anything but new. It was already implicit in Rachel Carson's *Silent Spring* (1962) which first alerted the world to the danger of human civilization killing off nature (in her case the excessive use of DDT), and which is widely considered as the starting point of the contemporary ecological debate. In the literature of the environmental movement since the 1970s the idea is known, for example, through Edward Goldsmith. In a text first published in 1977 and republished 11 years later, he voiced his concern that a 'surrogate world' would gradually replace the 'real' one. It was 'the technosphere or world of material goods and technological devices' (1988, p. 185) which Goldsmith called the surrogate world, and for him 'the process of building (it) up' implied the 'contraction and deterioration of the real one' (ibid., p. 187). In his simple model, the artificial world was pushing the other back spatially. Yet the real world, nature, would always be more real than the surrogate world, and, although it would have followed logically, it was inconceivable to Goldsmith that it might lose its reality and be entirely replaced by the surrogate world. In 1980 Carolyn Merchant announced *The Death of Nature*. Merchant tells the story of modern science and its tool, instrumental reason, which subjugates nature and denies the existence of the non-rational. By and large she reiterates the criticism voiced so powerfully in Horkheimer and Adorno's *Dialectic of Enlightenment* (1944/1972) and further developed in the later works of the Frankfurt School of Critical Theory.

While Merchant only pointed towards the possibility and danger of the future abolition of nature, Bill McKibben, in *The End of Nature* (1990), tried to take its historical reality as the starting point for his considerations. Acid rain and human induced climatic change, McKibben argues, have converted every spot of pristine nature into human envitronment. The planet we are inhabiting is therefore a 'post-natural world' (p. 55), and his book is supposed to be written from a post-natural perspective. McKibben is terrified by the idea that mankind has taken on the task of global management, not just because he believes that human managerial competence and strategies will never become sufficiently sophisticated to match the complexity of the task. For him, an equally significant dimension of the catastrophe is that after the end of nature 'there is nothing but us' (p. 55). Living in a 'post-natural' world means that 'we can no longer imagine that we are part of something larger than ourselves' (p. 78). What he refers to is the cultural significance of nature as the paragon of the good, true and beautiful which he believes to be indispensable as the ultimate source of human values and the meaning of human life. Here, McKibben touched upon a fundamental point, yet he did not spell out its implications. Irrespective of his own diagnosis according to which the patient was already dead, McKibben still thought it useful to devise a therapy. From the post-natural world he fell back into the appellative mode known from the earlier literature, and formulated imperatives to secure nature's survival. A further belief McKibben shared with Goldsmith and Merchant is that the end of nature—historical fact or still to take place—is an entirely and exclusively negative caesura.

Around the turn to the 1990s, however, this pessimistic, loss-oriented outlook was complemented by much more positive views. In particular Ulrich Beck's *Ecological Politics in the Age of Risk* (1988/1995a) and *The Re-invention of Politics* (1993/1997) strike an entirely different cord. These books try to differentiate between the two dimensions of nature which McKibben confused, but they also demonstrate that one can hardly be thought without the other. In a seemingly conventional way Beck argues:

> The process of interaction with nature has consumed it, abolished it, and transformed it into a civilisatory[1] meta-reality that can no longer rid itself of the attributes of human (co-)creation.
>
> (Beck, 1995a, p. 37)

Here Beck is referring to the material nature, to 'that world entirely independent of us which was there before we arrived and which encircled and supported our human society' (McKibben, 1990, p. 82). But for Beck the real significance of nature lies in its other dimension, i.e. its function as a metaphysical source of meaning and a normative standard which is accredited to it because it 'was there before we arrived'. It really was *a priori*, i.e. prior to human experience. Arguably, the most fundamental concern of the eco-movement is the loss of nature as an absolute and indispensable normative standard for the organisation of social relations (Blühdorn, 1997). For Beck, however, the abolition of nature in this sense, and the invalidation of the related moral imperatives, is a positive development because, as he argues, whatever we call natural tends to be the product of naturalisation: 'after all, we decide who and what is nature, by means of science if need be' (Beck, 1997, p. 86). Beck urges us to acknowledge the 'irreversible artificiality of nature' (1995a, p. 37). He is concerned about the ideological potential of the issue of nature which he regards as a 'political chameleon' (1995b, p. 191). In a cultural context where reliable social values and binding ethical guidelines are high in demand but chronically low in supply, nature and the natural may serve as political justification for all kinds of ideological currents. Beck strongly agrees with Gernot Böhme who argues:

> The contemporary invocation of nature as a value proves ideological because it makes reference to the conception of nature as something unchanging just at the point of time when—historically probably irreversibly—it is disintegrating.
>
> (Böhme, 1992, p. 115)

In his 1992 book on nature in the age of its technical reproduction Böhme further explains:

> Making reference to nature as it is in itself . . . , making reference to human nature . . . , making reference to nature as a norm . . . ,

is an illusion which actually prevents the most fundamental problems from being addressed. In view of this situation one might begin to feel sympathetic to the demand to abandon the concept of nature altogether.

(1992, p. 22)

In order to prevent a possible anti-democratic and anti-modernist twist of ecological concerns; in order to avoid nature being instrumentalized for anti-ecological ideologies, Beck and Böhme demand 'to make nature itself political, i.e. to negotiate politically which nature we want at all' (Böhme, 1992, p. 24). The old backward oriented project of saving nature is wiped out by two fundamental insights: 'Firstly there is no movement seeking to protect the *Other* of society; secondly, there is no natural nature' (Beck, 1995b, p. 182). It is replaced by the future oriented project of democratically negotiating nature: 'nature becomes a societal project, a utopia' (Beck, 1997, p. 114). In actual fact, the abolition of nature becomes a political demand, an 'important and urgent' prerequisite for genuinely democratic politics. As nature can be used to legitimise just about any political ideology, only the democratization of nature can lead to 'an ecological democracy, which may control the anti-modern, dictatorial potential' of the ecological issue (Beck, 1995b, p. 191). Once again taking the philosophical and the physical dimensions of nature together, Beck describes the abolition of nature as the fusion between nature and society:

> *nature*, the great constant of the industrial epoch, is losing its preordained character, turning into a product, the integral, shapeable 'inner nature' of (in this sense) post-industrial society. . . . 'Nature' becomes a social project, a utopia that is to be reconstructed, shaped and transformed. *Re*naturalization means *de*naturalization. Here modernity's claim to arrange things has been perfected under the banner of nature. Nature becomes politics. . . . That means that society and nature fuse into a 'natural society', either by nature becoming societalized or by society becoming naturalized.
>
> (Beck, 1997, pp. 114f)

For the eco-movement and ecological thought this fusion between nature and society has significant implications. As Beck notes, 'the social "consumption" of nature renders philosophically invalid all those conceptions and theories of nature which conceive of it as the counter-image of human activity and power' (1995a, p. 38). This dissolves the normative foundation of ecological claims, and the eco-movement as well as their theorists are in danger of falling prey to a *naturalistic fallacy*:

> In the context of the ecological debate, too, the attempts to use nature as a measure against its own destruction are subject to a self-misunderstanding: the nature to which they refer in order to stop its destruction, does no longer exist.
>
> (Beck, 1995a, p. 62)[3]

Anthony Giddens who points to the close connection between the *natural* and the *traditional*, and suggests that 'the end of nature—as the natural—coincides with the end of tradition—as the traditional' (Giddens, 1994, pp. 85f), highlights the same problem:

> Decisions about what to conserve, or to strive to recover, can rarely be decided by reference to what exists independently of human beings.
>
> (Giddens, 1994, p. 102)

Without this external point of reference, however, the environmental discourse becomes disoriented. After the abolition of nature as the Other of society, which existed prior to and independent of society itself, the process of ecological modernization loses its sense of direction. A previously metaphysically meaningful (aimful) environmental debate is now struggling to define its own goals. It has to provide evidence that it can 'make sense' itself. Nevertheless, to the extent that the end of nature dismantles ideological conceptions of the natural and creates space for the emergence of social ideals and opportunities, it may still be regarded as a positive development. For ecological thought, the abolition of nature implies that rather than trying to enhance its ecocentric credentials, it has to become fully anthropocentric. In the past, the eco-movement and the theoretical writing accompanying it have largely been fuelled by the experience of the alienation from nature and the resulting desire for reconciliation with it. With the abolition of nature, this thinking in dualisms is abandoned. Neither alienation nor reconciliation will henceforth be possible. This shift from the paradigm of the loss of nature to that of its social creation demands that ecological thought turns away from the Other of society and transforms into social theory.

In two respects, however, any talk about the abolition of nature is dangerous and might give rise to fatal misunderstandings. In the interest of clarity it might therefore be useful to reiterate that, first, the abolition of nature does not mean that the concept of nature becomes irrelevant, or that political campaigning for it becomes obsolete. Secondly, it does not mean that human civilisation has become a closed system capable of reproducing itself without having to rely on external input. What has been abolished is *only*—and this is more than the European culture of enlightenment can easily digest—the notion of nature as the Other of society. But this must not be understood simply as the transformation of pristine nature into human environment. This would be a very limited understanding of the dissolution of the modernist dualism between nature and society. Much more significant is what we may call the discursive abolition of nature, i.e. the realization that nature as a normative standard, a quasi-metaphysical value, is no longer available because, in reality, the natural has never been more than the product of history and tradition which in a *post-natural* and *post-traditional* age (Giddens, 1994) is up for social renegotiation and reconstruction. Of course, post-natural societies, too, will require social values on whose basis they can establish systems of meaning. And as the ideas of nature and naturalness have by no means lost their intriguing attractiveness, Jagtenberg and McKie are correctly pointing out that 'for a corpse, nature maintains a vigorous afterlife' (1997, p. 21). Beyond the abolition of the traditional single nature which was assumed to be completely separate and independent from human civilization, nature survives as a multiplicity of competing social constructions.

III ENVIRONMENTAL SOCIOLOGY AND SOCIAL CONSTRUCTIONISM

To what extent have ecological theorists actually taken account of the abolition of nature? Up to the present, one of the major restricting factors of ecological theory has always been its subordination to ecological campaigning. As an immediate consequence, lines of enquiry which were suspected to be damaging rather than supporting the ecological cause were unlikely to be followed in any detail. Only with the emergence of environmental sociology as a by and large independent academic discipline, the study of the ecological debate itself, i.e. the mechanisms of its internal functioning and the analysis of its claims making procedures became possible. As we shall see in the next section, there is still significant pressure on ecological theorists to be ecologically correct. Nevertheless, in recent years the supposedly anti-ecological idea of the abolition of nature and the portrayal of nature as a social construct have gained currency. One major reason for this is that constructionist approaches have proven particularly useful for helping to explain how and why environmental concerns emerge, and how exactly such concerns translate into actual environmental policies.

While in earlier phases of the eco-movement the general assumption had been that by means of information and environmental education the problems 'only' have to be brought to the awareness of the wider public, and would then most certainly translate into remedial policies, it became increasingly clear in the 1990s that there is no direct connection between physical environmental change and political agendas. Ecological theorists like Yearley, Eder or Hannigan could build on the work of coustructivist pioneers like Kitsuse and Spector who had, already in the early 1980s, suggested that a clear distinction ought to be made between 'objective conditions and the subjective awareness of social problems' (1981, p. 199). Towards the end of the 1980s, for example the widely read books by Ulrich Beck popularized the observation that environmental 'destruction and protest are symbolically mediated' (1995a, p. 45). In the contemporary debate it is widely accepted that 'public concern about the environment is by no means automatic even when the conditions are visibly bad' (Hannigan, 1995, p. 2). Still, the idea that the 'perception of environmental problems may ... be independent of the magnitude of the problems themselves' (ibid., p. 24) remains rather irritating. Beck went as far as speaking of 'the law of independence of protest and destruction' (1995a, p. 45). If the social perception of problems is not directly related to objectively problematic conditions in the physical environment, the question arises 'whether the actual existence of the objective conditions is necessary before a social problem can arise' (Yearley, 1991, p. 49). If even 'visibly bad' conditions in the environment do not necessarily arouse social concerns; if, in other words, the history of the environmental movement is not simply the history of trying to force objective problems into the public consciousness and on the political agenda, it may, in turn, also be possible that environmental problems can be socially constructed and perceived even in the absence of hard evidence of ecological threats. The debate around the disused oil-platform Brent Spar might be a good case in point.

While it has remained impossible to determine the exact relationship between problems and concerns, it has become evident and uncontested that:

> public concern is at least partially independent of actual environmental deterioration and is shaped by other considerations; for example, the extent of mass media coverage.
>
> (Hannigan, 1995, p. 24)

Arguing along the same lines, but trying to reach even beyond the primary level at which actors like scientific experts, environmental pressure groups and the media construct the social perception of environmental problems, Ulrich Beck raises the attention to

> the profound significance of the *cultural dispositions to perceive*, and of cultural norms. These decide *which despoliations are put up with and which are not*, and how acceptance of the unacceptable arises and persists against a background of unquestioned assumption.
>
> (Beck, 1995a, p. 45)[4]

So, if we want to know how environmental sociologists have responded to the abolition of nature, we can first of all note that they have begun to make a difference between empirically measurable environmental phenomena and the social perception of environmental problems. The recognition of the relative independence of environmental concern and environmental politics from actual physical environmental change has led to a shift of interest: away from the physical conditions towards the social mechanisms of problem construction and the 'cultural dispositions' on which they rely. This shift had already been demanded by Kitsuse and Spector who regarded the task 'to account for the emergence and maintenance of claim-making and responding activities' as the 'central problem' for further sociological research (1981, p. 201). Making explicit reference to their work, Steven Yearley suggested:

> that sociologists concerned with social problems should suspend any interest in whether the objective circumstances merit the existence of a social problem or not. . . . Instead they should focus on the social processes involved in bringing an issue to public attention as a social problem.
>
> (Yearley, 1991, p. 50)

Or as John Hannigan phrased it a few years later:

> From a sociological point of view, the chief task here is to under-
> stand why certain conditions come to be perceived as problematic
> and how those who register this 'claim' command political atten-
> tion in their quest to do something positive.
>
> (Hannigan, 1995, pp. 2f)

This change in the research agenda 'encourages us to examine
processes *internal* to the green movement' (Yearley, 1991, p. 52; my
emphasis). This new inward orientation reflects the abolition of nature
as the Other of society, and implements the demanded turn of ecological thought into social theory. It leads to a reinterpretation of the
eco-movement as a 'society-oriented, inward movement' rather than
an 'outward-oriented movement for the environment'[5] (Beck). Such
an inverted approach to the ecological issue gives rise to the challenging thesis that:

> It is not the despoliation of nature, but the jeopardization of a *spe-
> cific cultural model of nature* . . . that provides the sounding-board
> for the ecological alarm of an entire society.
>
> (Beck, 1995a, p. 54)[6]

John Hannigan has identified three major advantages of a social constructionist approach in environmental sociology. First, such an
approach is more critical than much of the traditional environmental
literature in that it does not simply 'accept the existence of an environmental crisis brought on by unchecked population growth, over-
production, dangerous new technologies' (Hannigan, 1995, p. 30).
Instead of reproducing the stereotypes of the eco-debate, it focuses on
'the social, political and cultural processes by which environmental
conditions are defined as being unacceptably risky and therefore
actionable' (ibid.). Indeed, a constructionist approach can contribute
significantly, for example, to the struggle against anti-modernist
and evidently eurocentric interpretations of the global environmental crisis. Secondly, a social constructionist approach 'recognises
the extent to which environmental problems and solutions are end-
products of a dynamic social process of definition, negotiation and
legitimation both in public and private settings' (ibid., p. 31). And
finally, by deflecting the attention from the so-called objective external
conditions which are traditionally the realm of the natural rather than

the social sciences, 'a social constructionist approach grounds the study of environmental matters in a distinctly sociological discourse' (ibid.). Certainly environmental sociologists working with a constructionist model could never pretend to be able to reveal the full and ultimate truth about ecological issues, but it is evident that their specific perspective reveals insights which remain inaccessible to both natural scientists and eco-campaigners. After all, it questions what both of these uncritically presuppose: nature and the environmental problem.

IV   BULLDOZERS AND CHAINSAWS

Although there is now a general consensus that both nature and ecological problems are at least in part social constructions which are subject to a continuous process of discursive deconstruction and reconstruction, social constructionists still have to face powerful opposition from those refusing to accept the abolition of nature. Environmental realists like Benton, Dunlop and Catton, Martell, and to some extent even Ulrich Beck himself, insist on the independent existence of nature and the objective reality of environmental problems without which it would in their view 'be difficult' for the social constructionists 'to define such problems as existing or arouse popular concern over them' (Martell, 1994, p. 132). Soulé heavily attacks 'the social siege of nature' (Soulé and Lease 1995). Martell warns that 'the objective basis of environmentalism in problems in the environment can be sociologized away' (ibid., p. 132). He urges social constructionists to make use of soiology's 'chance in its encounters with the non-human world to be more naturalistic and objectivist and less socially reductionist in its understandings' (ibid., p. 13). Ted Benton voices similar concerns when he argues against an 'oversocialised view of nature' (Benton, 1994, p. 45). According to Benton, social constructionists favour 'a perspective in which the independent presence of the non-human world in our lives is marginal to the point of disappearance' (ibid.). Making explicit reference to Yearley's social problems perspective and Yearley's emphasis on analysing the societal processes of problem construction, Benton argues:

> It is, of course, both interesting and important to be aware of these social processes, but in the case of environmental issues this approach has the consequence of bracketing out of sociological analysis any consideration of the 'objective conditions' which give

rise to environmental concern. It is all the same, as far as the sociologist is concerned, whether we do, in fact, face ecological catastrophe, or whether environmentalists have conjured this threat out of their fevered imaginations.

(Benton, 1994, p. 46)

Arguing along the same lines, Dunlap and Catton insist:

treating global environmental change . . . as a social construction discourages investigation of the societal causes, consequences and amelioration of global environmental problems . . . this seems particularly unwise in the case of global environmental change.

(1994, p. 20)

Attacks against the social constructionist approach tend to become particularly serious when constructionist thinking is seen to be inspired by elements of postmodernist thinking. It is evident that the abolition of the singular nature of modernity and its reconceptualization as a social construct clear the way to the post-modern pluralization of nature. Discursive deliberations about nature are most likely to generate multiple and competing concepts of nature and the natural. This however is feared to undermine 'the objective basis for moral concern' (Martell, 1994, p. 132) about environmental issues. It therefore comes as no surprise that 'for the most part, environmental sociologists have eschewed involvement in this debate (modernity versus postmodernity), preferring to deal with more empirically grounded research problems' (Hannigan, 1995, p. 178). As Hannigan correctly points out, 'most environmental researchers who have considered the matter have shied away from adopting a postmodernist perspective' (ibid., p. 181). Instead, environmental realists have identified 'postmodern deconstructionism' as a new political enemy, which is itself regarded as part of the ecological problem.

As if the abolition of nature could be stopped or reversed by militating against certain theoretical models proposed by sociologists, postmodernist deconstructionism is denounced as a 'contemporary form of intellectual and social relativism' that 'can be just as destructive to nature as bulldozers and chain saws' (Soulé and Lease, 1995, p. xvi). According to Soulé, postmodernist deconstructionism is a coalition of 'conservative free market capitalists, humanists concerned with the emancipation and empowerment of certain social and ethnic groups, and others, including animal rights organisations' (Soulé, 1995,

p. 146). Supposedly, these actors are all united in the attempt to 'justify further exploitative tinkering with what little remains of wildness' (ibid., p. xv). Postmodernist deconstructionism is assumed to be driven by the 'belief that a world beyond our control is so terrifying that we can—indeed, must—believe only in the landscapes of our imagination' (Shepard, 1995, pp. 22f). Whilst the environmental movement is seen as 'essentially a reawakening to the truth ... that we must depend on other forms of life to survive' (Worster, 1995, p. 79), 'deconstructionist postmodernism rationalizes the final step away from connection: beyond relativism to denial' (Shepard, 1995, p. 25). If we believe Shepard, it effectively means taking 'refuge from overwhelming problems by announcing all lands to be illusionary' (ibid.).

Invariably, such a brief overview of the criticism of social constructionist approaches fails to take account of the specific points and positions of individual environmental realists. Undeniably, there are different degrees of realism, just as we may differentiate between mild and more radical forms of constructionism. Nevertheless, even such a rough sketch helps to get the main points across. The central message of environmental realism is 'that the world, including its living components, really does exist apart from humanity's perceptions and beliefs about it' (Soulé and Lease, 1995, p. xv). The ever repeated warning directed against the 'idealist tendency in modern social theory' is that the 'reality external to discourse' must not be reduced to 'an unknowable ghostly presence' (Benton, 1994, p. 45). And wherever social constructionism fuses with postmodernist thinking, this immediately rises the spectre of moral relativism: If all truth claims have validity, then there is no basis for endorsing some over others, and thus no basis for becoming proactive' (Dunlap and Catton, 1994, p. 22). Postmodernist social constructionists, in particular, are accused (i) of undermining the possibility to take a critical stance *vis-à-vis* ecologically damaging societal practices, (ii) of impairing the moral credibility of ecological campaigns, and (iii) of promoting political quietism by destroying the motivational basis for ecological action. Contrary to 'the subjective and aesthetic dandyism of our time' (Shepard, 1995, p. 27), environmental realists are convinced that 'the genuinely innovative direction ... is not the final surrender to the anomie of meaninglessness or the escape to fantasylands but in the opposite direction—toward affirmation and continuity with something beyond representation' (Shepard, 1995, p. 25). Whilst it will hardly be possible to deny that the realist concerns are at least partly legitimate, it is also obvious that there are serious errors and misconceptions in the realist position which need to be rectified.

V IN DEFENCE OF THE CONSTRUCTIONIST APPROACH

There firstly is the bizarre idea that social constructionists are aiming to 'justify further exploitative tinkering' with the natural environment and that they themselves are responsible for the destruction of nature. Admittedly, the constructionist project implies the deconstruction of certainties which have so far provided the basis for the ecological discourse. However, the certainties they question cannot be said to have been a particularly adequate foundation for environmental politics. There is little evidence suggesting that traditional environmental politics, and in particular ecological modernisation, has been very successful in terms of reaching the objectives it had formulated for itself. A major reason for the so far rather disappointing record of environmental protection was probably the lack of theoretical enquiry into the kind of nature environmentalists wanted to preserve, protect or restore. It is hence probably in the interest of a more successful politics of nature to deconstruct a concept of nature which has proven inadequate and thereby lay the theoretical foundations for more appropriate political strategies. Beyond this, it is hardly fair to say that environmental constructionists themselves are destroying these traditional certainties, when in fact they merely theorize and conceptualize societal processes which evolve and take effect quite irrespective of their being observed and theorized. What was described above as the abolition of nature is not the work of environmental sociologists, but their constructivist approach rather has to be considered as sociology's attempt to respond to it.

Secondly, it is a fundamental misunderstanding if social constructionists are seen to be terrified by the existence of a world which is beyond their control, and that they seek to perfect modernity's claim to absolute power by denying the existence of anything uncontrollable. If constructionists take a social problems perspective, and emphasize that processes of political agenda setting and environmental policy making respond first and foremost to socially constructed concerns rather than to the so-called objective realities, this certainly enhances rather than reduces the complexity of ecological issues. Particularly in their postmodernist variety, constructionists acknowledge that after the abolition of the great constant of modernity, i.e. the single nature as the Other of society, we are confronted with the emergence of multiple and equally legitimate representations of nature, which can no longer be forced under one unifying umbrella. It is therefore difficult to see how constructionist approaches can be said to increase human power and control. As some realist critics have correctly pointed out, the opposite is much more likely to be the case.

Thirdly, we have to consider the allegation that constructionists are denying the reality of overwhelming environmental problems and take refuge in the virtual reality of their imagination. If constructionists were to deny, firstly, that the physical environment is subject to processes of change which are accelerated by the impact of human civilization, and secondly, that contemporary societies devise concrete policies in response to these changes, they would certainly not deserve to be taken seriously. To what extent, however, these undeniable physical changes can be described as environmental problems and remedies, is a completely separate issue. Here we are confronted with the question of where and how contemporary societies find their measure or normative standard for both environmental deterioration and remedial action. Traditional environmentalism, although far from simply equating environmental change and environmental problems, has devoted amazingly little thought to the distinction between the two (Blühdorn, 1997). We can safely assume that nobody would identify environmental problems where change has not been classified as deterioration, but any such classification invariably relies on a set of norms and parameters which are socially mediated. Hence, environmental problems, unlike physical environmental change, are always and necessarily social constructions, which may legitimately be deconstructed and tested for their validity. And those engaging in this complex activity can hardly be accused of wanting to escape from reality into a fantasyland. On the contrary, social constructionists seek to face the realities we have constructed.

This consideration takes us to the central question of whether the abolition of nature and the resulting constructivist approach may imply the discursive abolition of the environmental problem. If nature and the environmental problem are social constructs and have no reality outside the societal discourse, the ecological crisis might turn out to be significantly less worrying than it had previously appeared. With regard to this question, social constructionists may be criticized for causing substantial confusion by failing to introduce clear terminological distinctions. This will be discussed in some detail in the concluding section, where we will also make suggestions how the constructivist approach could be improved. In direct response, however, to the realist allegation that constructionists reduce nature and the ecological problem to 'an unknowable ghostly presence', we can already at this stage offer some clarification: The whole debate about the independent reality of the physical environment outside the realm of social discourse may be an interesting question for epistemologists, in the context of environmental sociology, however, it is pretty irrelevant. Environmental constructionists may happily accept that the physical environment 'including its living components, really does exist apart from humanity's perceptions and beliefs about it'. Whether or not this is the case will neither be verifiable for them, nor is it particularly important as long as we realise that concrete environmental politics is not based upon such a hypothetical reality, but responds first and foremost to environmental concerns which are socially constructed by the means of science, certain interest groups, the media, etc. So, contrary to the realist criticism, the project of social constructionism is not at all concerned with denying or confirming *a priori* reality, but rather with exploring the composition and structure of the images on which modern societies base their politics, including their environmental politics. Whatever the external realities—for constructionist environmental sociologists the focus is on the question of how contemporary societies frame and process their *knowledge* about this external world. This exclusive focus on internal representations and their societal negotiation has in itself to be regarded as a significant and valuable contribution to the environmental debate.

VI TOWARDS A REFINED CONSTRUCTIONIST RESEARCH AGENDA

As we have seen in the previous section, the social constructionist approach can easily be defended against much of the realist criticism levelled against it. In particular, no realist insistence on the objective existence of nature and the ecological problem will reverse the cultural process we described as the abolition of nature. It remains a fact that there is no consensus about the nature we want to protect or restore, and that there is no normative standard which could guide our policy towards the environment. In other words, no realist criticism will remove the pressure on environmental sociology to devise theoretical models which take account of Hajer's 'new environmental conflict'. Of course this does not mean that the constructionist models which have been found are already fully satisfactory. Undeniably there are certain weaknesses which may be ironed out—yet not by returning to realist positions but by radicalizing the constructionist perspective. Where then do the constructionists go wrong, and how may their approach be improved? In this concluding section we want to argue that social constructionists themselves do not take the abolition of nature sufficiently seriously.

To begin with, we may note that Yearley, Beck, Hannigan and most other environmental sociologists who favour a social constructionist perspective are actually inconsistent in the way they handle the abolition of nature. When Beck, for example, argues that the environmental movement does not respond 'to the *despoliation* of nature, but the jeopardization of a specific cultural model of nature' (1995, p. 54; my emphasis), and that 'cultural dispositions' decide 'how acceptance of the *unacceptable* arises and persists' (1995, p. 45; my emphasis), he clearly accepts the extra-discursive 'objective' existence of 'despoliations' and 'unacceptabilities'. In other words, Beck presupposes a problem-standard which exists prior to and independent of the social discourse. Beck evidently preserves a conception of the normal and natural which has survived the abolition of nature as the Other of society. This is reflected in Beck's whole theory of the 'risk-society' (Beck, 1992) which assumes that late modern society is confronted with 'objectively' existing risks.

The same may be said about Hannigan who seems to distinguish between the '*perception* of environmental problems' and 'the magnitude of the *problems themselves*' (1995, p. 24; my emphasis), who notes that environmental concern 'is at least partially independent of *actual environmental deterioration*' (1995, p. 24; my emphasis), and who highlights that public problem perceptions do 'not necessarily reflect the *reality of actual problems*' (1995, p. 25). Anticipating the realist criticism, Hannigan even defensively points out:

> I am not by any means attracted to an extreme constructionist position which insists that the global ensemble of problems is purely a creation of the media (or science or ecological activists) with little basis in objective conditions. On the contrary . . . I fully recognise the mess which we have created in the atmosphere, the soil and the waterways.
>
> (Hannigan, 1995, p. 3)

Such defensive manoeuvres contribute more to confusing the situation than to clarifying it, for they leave uncertain what part of 'the mess' existed prior to the social discourse about it, and what the discourse constructed itself. Steven Yearley is slightly more cautious in his language, but when he distinguishes social problem perceptions from the 'objective conditions' or 'objective circumstances' (1991, p. 49), he effectively makes the same mistake as Beck and Hannigan. In other words, Yearley, too, seems to assume that there are 'objective' problems and risks which are ultimately more valid and real than the socially constructed risk perceptions. This is reflected in his suggestion that environmental sociologists should focus on the mechanisms by which a social problem 'comes to public attention', or 'comes to the fore' (1991, p. 50). It seems to be implicit that there are also problems which come neither 'to public attention' nor 'to the fore'. Yearley further insists, that social problem perceptions must have some kind of objective basis because 'whole groups of people do not get upset over "nothing"' (Yearley, 1991, p. 51). He explicitly endorses the position of Kitsuse and Spector, that 'group definitions of social problems *usually do* have reference to some empirically verifiable . . . objective condition' (Kitsuse and Spector, 1981, p. 200).

So what most social constructionists seem to share with their realist counterparts is the belief that there exist two clearly distinguishable worlds: the extra-discursive, empirically verifiable objective world, and the world of subjective impressions and socially constructed realities. Both these worlds are assumed to be accessible to scientific exploration. However, whilst the realists insist on investigating the *objective* conditions with the *real* problems as well as the *discursive* reality with the social problem *perceptions*, we have seen that constructionists believe they should 'suspend any interest in . . . the objective circumstances' (Yearley, 1991, p. 49), in order not to 'deflect attention from investigation of the definitional process' (Kitsuse and Spector, 1981, p. 200). For two reasons, however, the recommendation to leave the 'objective' conditions on one side, so that they may be explored by the natural sciences, while the social scientists deal with the subjective impressions and discursive constructions is ill-conceived. First, it is fundamentally wrong to believe that natural scientists could generate more objective truths than their colleagues in the social sciences. Even the so-called natural sciences will hardly be able to come up with anything but social constructions. And secondly, it is equally wrong to deny the subjective impressions and social constructions their full validity. To quote Yearley's example: 'people may get genuinely upset by the prospect of an invasion by wholly fictitious aliens' (Yearley, 1991, p. 52). Whatever their basis, these concerns are very real. And if they are shared by a sufficiently large number of people, they will undoubtedly have very real effects on political agendas. Hence, the recommendation to social constructionists should not be to leave the 'objective conditions' unobserved, but to take the social constructions seriously. Social constructions are the most real and most objective reality we have got, and nothing more real or more objective is accessible to us. This Kantian idea of the inaccessibility preserves the existence of an *a priori* world beyond the social constructions, yet it secures that the latter can claim full and unlimited validity. So the first suggestion for a refined constructionist research agenda is to take account of the full epistemological implications of the constructionist approach. The two equations *social = subjective/invalid*, and *natural = objective/valid* are both wrong. The same may be said of the opposition *social ≠ natural*. If the abolition of nature is taken seriously, the essentially modernist dualisms of nature/society and subjective/objective need to be given up or merged into the formula *natural = social = objective*. This does, of course, not mean to deny the ontological distinction between material and non-material. Neither does it dissolve the difference between 'social capital' (resources which can be socially produced and reproduced) and 'natural capital' (external input into social processes).

The reason why even constructionists hesitate to give up the belief in extra-discursive 'objective' problems is, firstly, that they are looking for an objective source of environmental concern—because people do indeed not get upset about nothing. Secondly, that health-hazards, resources shortages, the extinction of species and other issues which trigger such concern seem to be 'objective realities' outside the realm of societal discourse. With regard to the first point it may be said, that it is certainly correct to assume that environmental concerns need an 'objective' basis, yet there is no reason to believe that this basis necessarily needs to be located in the material world of physics. It may just as well be a non-material necessity originating from a certain (modern) way in which we conceive of ourselves. Taking the abolition of nature seriously means locating the objective basis of environmental concern on the side of the human subject and its self-conceptualisation as an identity. With regard to the second point it is worth reminding ourselves that it is never the material side of a phenomenon which triggers environmental concern, but always and exclusively the violation of cultural norms of naturalness and accept-ability. At an earlier point in this chapter we referred to these as the 'cultural dispositions to perceive' (Beck, 1995a). It is therefore the second recommendation for a refined constructionist research agenda, that environmental sociologists should look here rather than in the material world for the objective basis of environmental concern.

As these considerations suggest, social constructionists have so far not really started to explore the objective basis of environmental concern, erroneously believing that it is located in the extra-discursive material world. Yearly, Hannigan and other social constructionists have so far focused on science, pressure groups, the media and other 'moral entrepreneurs', and explored by what means these social actors turn environmental facts into political issues. However, if the emergence of social problem perceptions is exclusively explained as the work of 'moral entrepreneurs', this seriously overes-timates their power and influence. Apart from the fact that they frame and publicise phenomena of physical environmental change, these social actors are tapping a second resource which may be described as a pre-existing vital interest in nature—not as an accumulation of physical conditions, but as a metaphysical norm and context of values which is projected on these conditions. For the social construction of environmental problems this second resource is not just indispens-able, but it is even more important than the physical input. Problem perceptions may well correspond to empirically verifiable physical changes, yet this is by no means a necessary condition of their emer-gence. Hence the third suggestion for a refined constructionist research agenda is that environmental sociologists should not just focus on 'moral entrepreneurs' and the political career of single issues, but they should explore what is behind this pre-existing interest in nature or the standard of naturalness. Why do humans tend to get con-cerned about nature? Why do they care about certain physical phe-nomena which symbolize nature and about certain conditions which are perceived as unnatural? What does nature mean to us beyond the accumulation of physical conditions?

In particular when they are exploring how the ongoing evolution of modernity affects the need for and interest in this *meta*-physical nature, environmental sociologists are well advised to work together with philosophers and cultural theorists. Such co-operation may also help them implement the fourth and final recommendation for a refined constructionist research agenda: environmental sociolo-gists should overcome their reservations *vis-à-vis* postmodernist approaches. As we have argued above, it is no surprise that ecological theorists have so far shied away from postmodernist thinking. With its belief in the nature/society dichotomy ecological thought is fundamentally modern, and with its concept of the one all-embracing nature it is essentially anti-pluralist. Even less than with the idea that nature should be a social construct, ecological thought is compatible with the idea that there might be an irreducible plurality of *natures* (Blühdorn, 1997). The scepticism of environmental sociology *vis-à-vis* postmodernist thought, however, is a remnant of its subordination to ecological campaigning. In the interest of a more comprehensive understanding of the internal structure of the social discourse about ecology, environmental sociologists should fully emancipate themselves from their parent discipline and embrace the postmodernist paradigm of thought. Taking the abolition of nature seriously means taking account of its postmodern pluralization.

In conclusion to this chapter we may finally ask where such a refined constructionist research agenda is going to take us. In this context we once again have to remind ourselves that the prime interest of environmental sociology is not to *solve* the ecological crisis or to *save* nature. Environmental sociology is not meant to be ideological and politically prescriptive. Instead, it seeks to be analytical and understand what exactly this crisis consists in, what the concept of nature means to us, and why we want to save nature at all. So while we should better not expect environmental sociologists to devise more effective strategies for the protection of nature, we may well expect them to help us understand why environmental politics and in particular the strategy of ecological modernisation have so far had a rather modest record of success. If we want to understand how and why ecological claims are made, sustained, and (not) translated into actual policies, a constructionist approach is certainly much more promising than its realist alternative. But only a radicalized constructionist perspective which takes full account of the abolition of nature will help to show that in late modern societies the 'cultural dispositions to perceive' physical environmental change have become radically different. As a hypothesis to be investigated by further research we might suggest that late modern societies are running out of the central resource for environmental politics: the demand for the metaphysical norm and standard of naturalness is in decline. Multiple, socially negotiated and often mutually exclusive value systems have taken over the function formerly fulfilled by the idea of an all-inclusive nature. A radicalized constructionist approach is likely to reveal that the interest in nature is itself a (specifically modernist) social construction. Its postmodernist deconstruction may well lead to the realization that the idea of ecology as a new ideological master-frame integrating and serving the whole of humanity is an illusion. The theory of ecological politics will probably have to make room for a *theory of post-ecologist politics* (Blühdorn, 1997, 1999). An integrated global policy of ecological modernization will presumably turn out to be a structural misconception.

Notes

1. Official translation: 'civilizing'.
2. Omitted in the official translation.
3. Original emphasis dropped in the translation.
4. The official translation underemphasizes Beck's crucial point: 'The ecological movement is not an environmental movement but a social, inward movement' (Beck, 1995a, p. 55).
5. Original emphasis dropped in the translation.

Further Reading

U. BECK (1997) *The Reinvention of Politics: Rethinking Modernity in the Global Social Order*, Cambridge: Polity Press (translation of *Die Erfindung des Politischen*, Frankfurt: Suhrkamp, 1993).
K. EDER (1996) *The Social Construction of Nature*, London: Sage.
J. HANNIGAN (1995) *Environmental Sociology: a Social Constructionist Perspective*, London/New York: Routledge.
M. E. SOULÉ and G. LEASE (Eds) (1995) *Reinventing Nature? Responses to Postmodern Deconstructionism*, Washington, DC/Covelo, California: Island Press.

References

U. BECK (1995a) *Ecological Politics in an Age of Risk*, Cambridge: Polity Press (translation of *Gegengifte*, Frankfurt: Suhrkamp 1988).
U. BECK (1995b) *Die feindlose Demokratie. Ausgewählte Aufsätze*, Stuttgart: Reclam.
U. BECK (1997) *The Reinvention of Politics: Rethinking Modernity in the Global Social Order*, Cambridge: Polity Press (translation of *Die Erfindung des Politischen*, Frankfurt: Suhrkamp, 1993).
T. BENTON (1994) 'Biology and social theory in the environmental debate', In: Redclift and Benton (eds), 1994, pp. 28–50.

I. BLÜHDORN (1997) 'A theory of post-ecologist politics', In: *Environmental Politics*, vol. 6, no. 3, Autumn 1997, pp. 125–47.

I. BLÜHDORN (1999) 'Ecological modernisation and post-ecologist politics', in Buttel, F. and Mol, A. (eds) (1999) *Ecological Rationality and Modernity*, London/Thousand Oaks/New Delhi: Sage (forthcoming).

G. BÖHME (1992) *Natürlich Natur: über Natur im Zeitalter ihrer technischen Reproduzierbarkeit*, Frankfurt: Suhrkamp.

R. CARSON (1962) *Silent Spring*, London: Penguin.

P. CHRISTOFF (1996) 'Ecological modernisation, ecological modernities', *Environmental Politics*, vol. 5, no. 3, Autumn 1996, pp. 476–500.

R. DUNLOP and W. CATTON (1994) 'Struggling with human exemptionalism: the rise, decline and revitalization of environmental sociology', *The American Sociologist*, Spring 1994, pp. 5–30.

K. EDER (1996) *The Social Construction of Nature*, London: Sage.

A. GIDDENS (1994) *Beyond Left and Right: the Future of Radical Politics*, Cambridge: Polity Press.

E. GOLDSMITH (1988) *The Great U-Turn: De-industrializing Society*. Bideford: Green Books.

M. A. HAJER (1995) *The Politics of Environmental Discourse: Ecological Modernisation and the Policy Process*, Oxford: Clarendon Press.

M. A. HAJER (1996) 'Ecological modernisation as cultural politics', in Lash, S., Szerszynski, B., Wynne, B., (eds) (1996) *Risk, Environment and Modernity: Towards a New Ecology*, London/Thousand Oaks/New Delhi: Sage. pp. 246–68.

J. HANNIGAN (1995) *Environmental Sociology: a Social Constructionist Perspective*, London/New York: Routledge.

M. HORKHEIMER and T. W. ADORNO (1972), *Dialectic of Enlightenment* (translated by John Cumming), New York: Herder and Herder.

J. HUBER (1993) 'Ökologische Modernisierung: Zwischen bürokratischem und zivilgesellschaftlichem Handeln', in *Volker v. Prittwitz*, 1993, pp. 51–69.

T. JAGTENBERG and D. MCKIE (1997) *Eco-Impacts and the Greening of Postmodernity*, London/Thousand Oaks/New Delhi: Sage.

M. JÄNICKE (1993) 'Ökologische und politische Modernisierung in entwickelten Industriegesellschaften', in *Volker v. Prittwitz*, 1993, pp. 15–30.

J. KITSUSE and M. SPECTOR (1981) 'The labeling of social problems' in Rubington and Weinberg, 1981, pp. 198–206.

L. MARTELL (1994) *Ecology and Society: an Introduction*, Cambridge: Polity Press.

B. MCKIBBEN (1990) *The End of Nature*, London: Penguin.

C. MERCHANT (1980), *The Death of Nature. Women, Ecology and the Scientific Revolution*, San Francisco: Harper and Row

A. MOL (1996) 'Ecological modernisation and institutional reflexivity: environmental reform in the late modern age', in *Environmental Politics*, vol. 5, no. 2, Summer 1996, pp. 302–23.

V. VAN PRITTWITZ (ed.) (1993) *Umweltpolitik als Modernisierungsprozeß: Politikwissenschaftliche Umwelt-forschung und-lehre in der Bundesrepublik*, Opladen: Leske and Budrich.

M. REDCLIFT and T. BENTON (eds) (1994) *Social Theory and the Global Environment*, London/New York: Routledge.

E. RUBINGTON and M. WEINBERG (eds) (1981) *The Study of Social Problems*, New York: Oxford University Press.

P. SHEPARD (1995) 'Virtually hunting reality in the forests of Simulacra', in Soulé and Lease, 1995, pp. 17–29.

M. E. SOULÉ and G. LEASE (eds) (1995) *Reinventing Nature? Responses to Postmodern Deconstrutionism*, Washington, DC/Covelo, California: Island Press.

G. SPAARGAREN (1996) 'Ecological modernization theory and the changing discourse on environment and modernity' (unpublished), paper presented at the Euroconference on 'Environment and Innovation', Vienna, 23–26 Oct. 1996.

D. WORSTER (1995) 'Nature and the disorder of history', in Soulé and Lease, 1995, pp. 65–85.

S. YEARLEY (1991) *The Green Case: A Sociology of Environmental Issues, Arguments and Politics*, London: HarperCollins.

12 Forest Restoration Research in Conservation Areas in Northern Thailand

David Blakesley, J. Allister McGregor and Steven Elliott

Summary

Loss of forests and their associated biodiversity is a serious issue in many tropical countries. In Thailand, for example, forest cover has been reduced from about 53% in the early 1960s (Bhumibamon, 1986) to about 22.8% or 111,010 km^2 today (FAO, 1997). Until very recently, rapid economic development of the country and its integration into the global economy has been a major underlying cause of deforestation. Especially in the late 1970s and early 80s, deforestation was largely the result of logging and the expansion of agricultural land. Consequently today, secondary forests and completely denuded land occupy large areas. Although Thailand has an extensive system of national parks and wildlife sanctuaries, occupying about 13% of its area, even these so-called protected areas contain large deforested areas.

To counteract continuing deforestation, conversion of deforested areas back into forest is necessary to protect watersheds and conserve biodiversity. Large-scale restoration of forest ecosystems will require close co-operation between government agencies and local people; the development of new and technically sound methods of tree propagation and planting and the provision of resources and expertise to all participating organizations. Community forestry could give local people greater control over local forest resources, but such programmes will also require technically sound restoration strategies if they are to provide the required diverse range of forest products and ecological services.

In 1993 a major reforestation project was initiated in Thailand to mark His Majesty King Bhumibol Adulyadej's Golden Jubilee. The long term aim of this project is to plant native tree species on 8,000 km² of degraded forest land. In contrast to previous reforestation programs in Thailand, which involved planting monocultures of pines and eucalypts, the Golden Jubilee project is using a wide range of native forest tree species. Tree planting is being undertaken by a diverse range of organizations, including villages, schools, charitable foundations, religious groups, other NGOs and large corporations, such as the state oil and electricity companies. The project will continue until 2002.

Nursery practices in Thailand are very advanced for woody species, but this technology has concentrated almost exclusively on commercial plantation species. Knowledge of the habitat requirements and how to propagate and plant the vast majority of Thailand's estimated 3,600 native tree species is very limited. Recognizing the need to initiate research on the scientific and technical aspects of forest restoration, Chiang Mai University and the Royal Forest Department jointly founded the Forest Restoration Research Unit (FORRU) in 1994, in collaboration with the University of Bath and the Natural History Museum, London. Its aims are to develop tools for studying the restoration of natural forest ecosystems; to determine ways in which these processes might be accelerated, to develop methods to propagate appropriate tree species for experimental planting trials and to provide information to organizations and individuals involved in reforestation programs. To achieve this, FORRU is building relationships with local villagers and educational organizations to ensure that its scientific resources are made available to people carrying out tree planting projects. Such relationships ensure that appropriate research is undertaken and that appropriate educational materials are produced. A successful outcome to this project will be the adoption of effective and appropriate nursery practices and the cultivation of a wide range of native tree species by organizations carrying out reforestation projects.

I INTRODUCTION

Deforestation in the Region

Southeast Asia, like most tropical regions, is losing its forests and associated biodiversity at an alarming rate. In Thailand, for example, forest cover has been reduced from about 53% in the early 1960s (Bhumibamon, 1986) to about 22.8% or 111,010 km² today (FAO, 1997). Unofficial estimates, however, put Thailand's natural forest cover at less than 20% (Leungaramsri and Rajesh, 1992). Thailand's rate of deforestation peaked in the late 1970s at about 6% per year and reached a minimum of about 0.3% per year in 1989, when the government banned commercial logging. Since then, however, the rate of forest destruction has increased to about 1% per year. Until very recently, the major cause of this considerable loss of biodiversity was poorly managed, rapid economic development and Thailand's increased integration into the global economy. Deforestation has largely been caused by logging and agricultural expansion (Hirsch, 1990). Consequently today, secondary forests subjected to differing degrees of disturbance and completely denuded land occupy large areas. Although Thailand has an extensive system of national parks and wildlife sanctuaries, occupying about 13% of its area, these so-called protected areas contain large deforested areas.

A positive development, however, has been a rapid increase in public awareness of the problems caused by deforestation in Thailand. It is now generally accepted that further loss of forest will cause more extreme floods and droughts, damage to watersheds, loss of biodiversity and impoverishment of rural communities (Elliott *et al.*, 1995). Economic and legal constraints continue to prevent complete protection of all remaining primary forest. Therefore, to counteract continuing destruction, deforested areas must be converted back into forest.

Following the ban on commercial logging in 1989, the National Forest Policy, which stipulates that 40% of the country should be under forest cover, was adjusted. The target for production forest was reduced from 25% to 15% of the country's area, whilst that for conservation forests was increased from 15% to 25%. This policy was implemented by designating many former logging concessions as national parks or wildlife sanctuaries. Such areas now cover about 72,020 km² (13% of the country, or more than half of the total forest area (Boontawee *et al.*, 1995). Consequently large parts of many national parks and wildlife sanctuaries were already degraded or deforested before they acquired protected status. If such areas are to fulfil their functions of conserving biodiversity and protecting watersheds, they must be reforested.

Forest Types in Northern Thailand

Most of the lowlands of northern Thailand were originally covered by deciduous forest dominated by teak (*Tectona grandis* L.f. (Verbenaceae)). Extensive logging, mostly by European timber companies, in the late eighteenth to early nineteenth centuries removed most of the teak, so that today this original forest type is found in very few areas (e.g. Mae Yom National Park). Instead, bamboo-deciduous forest has replaced the former teak forests, whilst in more severely degraded areas, deciduous dipterocarp-oak forest predominates. At mid elevations (about 600–1,000 m) a mixed deciduous forest occurs with very high tree species richness, whilst above about 1,000 m elevation, the forest is evergreen. In areas affected by fire, deciduous trees grow at higher elevations than usual, whilst along permanently wet stream valleys, evergreen trees occur at lower than normal elevations (Maxwell *et al.*, 1995, 1997).

Logging, shifting cultivation and infrastructure development projects are continuing to destroy all forest types in northern Thailand. Populations of large vertebrates have been severely depleted and many species have become extirpated from the region. Consequently, biodiversity in northern Thailand, which is extremely rich but poorly known, is now severely threatened. The Government of Thailand has recognized these problems and set a target to reforest 72,000 km^2 under the 6th, 7th and 8th National Economic and Social Development Plans.

Within national parks an wildlife sanctuaries, where the primary objectives are wildlife conservation and watershed protection, reforestation efforts should aim to permanently restore original forest ecosystems, as closely as possible, by accelerating the natural processes of forest regeneration. Rehabilitation and restoration of degraded ecosystems are extremely important components of *in situ* conservation as identified in the Convention on Biodiversity (World Conservation Monitoring Centre, 1996), but research on the biological and social dimensions of these issues has been neglected. Large-scale restoration of forest ecosystems requires close co-operation between government agencies and local people; the development of new and technically sound methods of tree propagation and planting and the provision of resources and expertise to all participating organizations. The Thai government is currently formulating a community forestry law that would give local people greater control over local forest resources, but enactment of such legislation will also require technically sound restoration strategies, if community forests are to provide the required diverse range of forest products and ecological services.

Requirement of various groups for tree nursery and propagation technology

In the past few years there has been a major shift in the types of reforestation projects being carried out in Thailand, away from the establishment of plantations of fast-growing species such as pines and eucalypts, towards planting a wide range of native forest tree species. This type of reforestation is sometimes termed 'enrichment planting' or 'accelerated/assisted natural regeneration'.

The major reforestation project initiated to mark His Majesty King Bhumibol Adulyadej's Golden Jubilee had the long-term aim of planting native tree species on more than 8,000 km^2 of degraded forest land. Tree planting is being undertaken by a diverse range of organizations, including villages, schools, charitable foundations, private companies, religious groups and other NGOs. The enthusiasm with which local people have participated in the first tree planting events demonstrated immense public support for forest restoration on a large scale (Elliott *et al.*, 1995). In terms of providing people with the opportunity to get directly involved in environmental protection and improvement, these projects are clearly successful. However, the project has faltered due to the current economic crisis and the setting of over-ambitious planting targets. Poor species selection, lack of after care of seedling plantings, drought and especially rampant fires killed many of the seedlings planted at the start of the project. Figures for mortality are unavailable, since there was very little quantitative monitoring of seedlings after planting, but losses are considered to be at least 40% and possibly more (Sukpanich, 1998). Guidelines to maximize the effectiveness of these efforts are urgently needed. There is no doubt that the effectiveness of forest restoration projects would be considerably improved through the development of simple, cost-effective, but scientifically sound tree propagation methods and the dissemination of this knowledge. The main groups requiring knowledge of nursery techniques are:

(1) Villagers and their supportive NGOs who want to include tree planting as part of community forest projects.
(2) Royal Forest Department officials restoring protected areas.

(3) Groups involved in the Golden Jubilee project, including NGOs and private sector sponsors.

Current state of nursery technology in Thailand

Nursery technology for woody species in Thailand is quite advanced, both for conventional propagation and micro-propagation. However, this technology has concentrated almost exclusively on exotic and commercial plantation species. Much research on conventional propagation has been carried out at the ASEAN–Canada Forest Tree Seed Centre, Muak Lek, where, for example, cost-effective methods for the commercial propagation of tree seedlings have been developed using coconut husk (Kijkar, 1991a). The centre has also developed seed testing standards for commercial species such as *Dipterocarpus alatus, D. intricatus* and *Hopea oderata* (Krishnapillay, 1992) and vegetative propagation techniques for dipterocarps in general (Kantarli, 1993). Vegetative propagation of exotic species such as eucalypts and acacias has been extensively researched, resulting in the production of a handbook for the propagation of *Eucalyptus camaldulensis* (Kijkar, 1991b), for example. Elsewhere, researchers have developed micro-propagation techniques for bamboo (Gavinlertvatana, 1992). Also, plant biotechnology may offer some potential for the storage of seed and vegetative material of forest species through cryopreservation (Blakesley *et al.*, 1996)

However, very little work has been carried out at these institutes on the vast majority of Thailand's estimated 3,600 native forest tree species. Also, many of the conventional techniques developed for commercial species, such as the coconut husk medium would be inappropriate for many isolated small scale forest nurseries, and could not be directly transferred to such operations.

Research needed

Attempts to recreate natural forest ecosystems are hindered by their complexity. Any individual forest type may contain several hundred tree species, each of which may have evolved intricate relationships with hundreds of other organisms, such as herbivores, pollinators and seed dispersers. Restoration of natural forest ecosystems therefore requires a vast amount of ecological information, only a small fraction of which is currently known. We need to understand how forests regenerate naturally, identify the factors limiting regeneration and develop effective methods to counteract them and thus accelerate regeneration (Hardwick *et al.*, 1997). Planting nursery produced seedlings is just one of the many options; others include cultivation and husbandry of seeds, seedlings and saplings which are already present or preventing fire by maintaining a network of fire breaks. This approach is known as assisted or accelerated natural regeneration (ANR).

Whether or not seedlings are planted, it is essential to conserve any existing woody plants during site preparation and management. The timing and methods used to control weeds require particular attention. Indiscriminate slashing of weeds, to provide easy access to the site for tree planters, often destroys naturally established tree seedlings. In sites with high densities of naturally established seedlings, planting fails to replace the seedlings destroyed during site preparation. In such areas, tree planting actually results in a net decrease in tree seedling density. Merely slashing the above-ground parts of weeds, can promote further weed growth, especially for grasses, which deprives the planted seedlings of nutrients and moisture at the crucial time of establishment just after planting. If weeding is carried out by hand tools, it is therefore necessary to dig out the roots of the weeds. This is extremely labour intensive and, depending on labour costs, very expensive. An alternative is to use a non-residual herbicide, such as glyphosate. This requires far less labour and is much cheaper than using hand tools. However, it is difficult to avoid spraying both naturally established and planted tree seedlings and can result in very high seedling mortality. Potential safety issues and the risk of pollution accidents also have to be considered, especially when working with untrained volunteers. In addition to the method of weeding, its timing also requires further research. While weeds almost certainly compete with tree seedlings for light and nutrients during the rainy season, when moisture is plentiful and growth is rapid, they might actually shade seedlings from excessive heat during the dry season. Tree seedlings of different species may respond to the various herbaceous weed species in different ways, so that weeding would be beneficial to some tree seedling species and harmful to others. It is even possible that seedlings of different ages or sizes are affected by weeds differently. The scope for research to improve site preparation and weeding methods for tree planting projects is clearly considerable.

Species selection is also important. The soil and climate requirements of commercial timber species are well known, but for the hundreds of other native forest tree species, very little is known. Sites vary enormously in soil, micro-climate, topography, etc. and each species has its own particular requirements. Without knowledge of such requirements, the selection of species for planting is very much a matter of chance. In addition to matching seedling species with prevailing site conditions, species chosen for planting should complement those already established naturally. Although planting additional seedlings of species already present (and therefore suited to local conditions) might result in a high survival rate and increase tree density, it would not increase the diversity of the regenerating forest. Within conservation areas, where increasing biodiversity is a major aim of tree planting, rapid survey techniques should be developed to determine which seedling species are already present. Species absent from, but otherwise suited to the site should then be chosen for planting. Such species are often those limited by inadequate seed dispersal. Uncontrolled hunting has eliminated many large bird and mammal species from some national parks and wildlife sanctuaries. Those trees which depend on large animals for seed dispersal may be unable to disperse their seeds into gaps, even though some of them may be able to grow under gap conditions. Research needs to be carried out to determine whether or not local extirpation of large birds and mammals is reducing the tree diversity of regenerating forests and to identify the tree species which may be affected. Species selected for planting should be fast-growing with dense spreading canopies, so that they can rapidly shade out weeds. They should also provide wildlife resources (such as fruit, nectar or perching sites) at a young age, so that animals (especially birds and bats) attracted by such resources, disperse the seeds of other non-planted tree species into the planted sites, thus accelerating the return of biodiversity. Such species have been termed 'framework' species (Lamb *et al.*, 1997) and research is urgently needed to discover which trees from among Thailand's native tree flora meet these criteria.

One major constraint to research on forest tree seedlings is the lack of an identification guide to seedlings. Most floras identify plant species on the basis of flower or fruit characteristics, so they cannot be used to identify seedlings. Identification of seedlings is crucial for many aspects of forest restoration work. To save time and nursery space, many seedlings used in current tree planting projects are dug up from remaining areas of forest and cultivated in a nursery for a year before being planted out. Such seedlings are often difficult to identify because the parent tree is unknown. To assess which species are regenerating naturally in deforested areas, it is necessary to identify seedlings of all species present at a wide range of sizes. Therefore, research on the morphology of seedlings, grown from seed of known parent trees is urgently needed to provide accurate descriptions for an identification guide to the forest tree seedlings of Thailand.

More research is also needed on the growing of seedlings in nurseries. Seeds should be collected from the nearest available seed source, so that the seedlings grown from them have a good chance of being genetically suited to local environmental conditions. Studies of the seasonal availability of seeds are needed to ensure that seed collection programmes in forest restoration nurseries are well planned. Seed germination also requires research. Although some tree species have a prolonged fruiting period, often the seeds within the fruits are viable for a very short period, and the stage of fruit ripeness may be critical. Many species have long periods of seed dormancy or low germination rates. For such species, treatments to break dormancy and improve the germination rate need to be devised. After germination, research must be carried out to determine the most effective conditions for growth, to ensure strong, healthy, high quality seedlings, which will be able to withstand the stress of transplantation into hot, sunny gaps. Such research would include experiments with different soils and other media, watering regimes, pest control and hardening-off treatments to prepare the seedlings for transplanting. After seedlings are planted out, further research is needed to develop appropriate husbandry regimes.

The enthusiasm generated by tree planting events is not always complemented by equal enthusiasm for caring for seedlings after planting and for monitoring their growth and survival. Considerably more research is required to provide the guidelines needed by the organizers and sponsors of tree planting programmes.

III   THE FOREST RESTORATION RESEARCH UNIT (FORRU)

Recognizing the need to initiate research into forest restoration techniques, The Forest Restoration Research Unit (FORRU) was established in 1994 to address some of the technical problems of re-establishing natural forest ecosystems on degraded sites within conservation areas (Elliott *et al.*, 1995). It is a joint initiative between Chiang Mai University (CMU) and the Headquarters of Doi Suthep-Pui National Park (under the Royal Thai Forest Department (RFD)) which adjoins the university campus and collaborates with expert advisors from Bath University and the Natural History Museum from the UK. The unit is situated near the Headquarters of Doi Suthep-Pui National Park (18°50′N, 98°50′E) at about 1,000 m elevation and has a permanent research staff of 5. It consists of a large nursery, an office and a computer room. The project has also recently established a community tree seedling nursery and field trials at Ban Mae Sa Mai, an Hmong hill tribe village in the north of the national park.

The aim of the project is to determine the most effective methods to complement and accelerate natural forest regeneration on deforested sites within conservation areas to increase biodiversity and protect watersheds. Specific objectives include:

(1) development of tools for studying the restoration of natural forest ecosystems, such as a seedling identification handbook, seedling herbarium and databases of seed, fruit and seedling morphology;

(2) understanding of the ecological processes of natural forest regeneration to determine ways in which these processes might be accelerated;

(3) identification of tree species suitable for planting to complement natural seedling establishment;

(4) development of appropriate methods to propagate such tree species and test their performance after planting out;

(5) training of interested groups in the new forest restoration techniques developed by the project

FORRU's primary task during the first phase of the project was simply to gather basic ecological information about the very large number of tree species which occur in northern Thailand, to determine which ones might be most useful for restoring damaged forest ecosystems. Research began on tree flowering and fruiting phenology and seed germination. Descriptions, drawings and photographs were made of fruits and seedlings. An herbarium collection of dried seedling specimens was established, along with computer databases of seed, fruit and seedling morphology. Without such basic background information, it was impossible to make sensible choices as to which tree species to use in forest restoration projects.

In addition, FORRU began to develop appropriate techniques for the propagation of about 40 native forest tree species, considered to be particularly suitable for forest restoration programmes. Seedlings are pricked-out, and grown on as container plants. From this work, it became clear that many species, such as *Hovenia dulcis* and *Sapindus rarak* germinate and grow readily in the nursery with no special treatments (Kopachon *et al.*, 1996). Other species, however, do not germinate readily or cannot be stored for even short periods (e.g. *Acronychia pendunculata, Craibiodendron stellatum, Rhus chinensis,* etc.). For these species, more detailed work is being carried out to determine appropriate treatments to overcome seed dormancy, etc.

In addition, the project has established experimental plots to assess the performance of seedling species after they are planted out in deforested plots. The plots are being established in co-operation with hill tribe villagers living within Doi Suthep-Pui National Park. FORRU has built a community tree nursery in the village of Mae Sa Mai, where the feasibility of the techniques developed at FORRU's research nursery is tested among local people. In addition to comparing seedling performance among species, the plots are being used to compare the effectiveness of different weed control, fertilizer and mulching treatments. The project is now concentrating on fast-growing species with dense, spreading crowns, which rapidly shade out weeds and those which provide wildlife resources (e.g. fruit, nectar perching places, etc.) which attract seed-dispersing animals into planted sites and accelerate the recovery of biodiversity.

IV   CONCLUSIONS

FORRU has only just begun to tackle the enormous research needs to improve forest restoration projects in northern Thailand. However, in the short time that it has been in operation, the unit has developed useful methods for studying a large number of native forest tree species (Elliott *et al.*, 1995). Such methods are probably applicable not only to Doi Suthep-Pui National Park, but to protected areas throughout the region. Data on individual species, however, are much more likely to be dependent on local conditions. Phenology, seed germination rates and seedling performance are probably all highly site-specific. Therefore a single research unit cannot complete the task. There needs to be a network of such units in various protected areas and effective exchange of information among them, so that regional recommendations as to the most effective methods of forest restoration can be developed. We hope that FORRU will motivate other workers to set up similar research units and encourage funding agencies to provide the necessary sponsorship. Ecological research can provide the most appropriate methods to restore natural forest ecosystems, but whether or not those methods are put into practice depends on social, political and economic factors. Research into these areas is therefore also essential. Given enough time and protection from disturbance, forests usually regenerate themselves. In areas where there is still an adequate seed source, tree cover can be restored within decades; in areas more distant from seed sources, it may take centuries. Whether or not this process should be accelerated will be decided by politicians, community leaders and land managers, not scientists; but in areas where a decision has already been taken to assist forest recovery, it is important to develop and use the most scientific and cost-effective methods available. Only more research can provide such methods.

V   TOWARDS AN ENVIRONMENT RESEARCH AGENDA

FORRU was established to address some of the technical problems of re-establishing natural forest ecosystems on degraded sites within conservation areas. As the project progressed, it became clear that connections should be made between the scientific resources developed at FORRU and organizations working with communities in Thailand on reforestation or smaller scale tree planting projects. Liaison with local groups and NGOs have ensured that appropriate research is undertaken, and that appropriate literature has, and will be produced. Further, FORRU expertise is being demonstrated and disseminated in ways suitable for a variety of end-users, including local communities and RFD personnel interested in planting native tree species. The output includes illustrated guides in local language, based on the scientific research, which will set out practical guidelines for the propagation of native species, including the construction and operation of appropriate low-tech nursery facilities. In order to achieve these objectives, contact has been made with NGOs and local communities in Thailand on reforestation or small-scale tree planting projects. For this type of project to be successful, consultation with local groups and NGO's must be continuous. It is also necessary to determine their current practices, and specific requirements for propagules of native species. Indigenous knowledge of the forest and of the socio-economic conditions as they impinge upon the forest is an important element, as it is necessary to learn from existing tree production efforts; what techniques are currently being employed, what special circumstances exist and their specific requirements for technology. It is also necessary to gain an insight into local factors such as the species of interest, propagation problems, local materials and finances available, and soil type availability. At the time of writing, the development of local capabilities to undertake tree planting and reforestation projects is continuing, and knowledge is being gained as to what species local communities regard as important, and would be selected by them in replanting programmes. Surveys of the socio-economic conditions of the communities in which the local groups and NGOs are operating could also be undertaken (Masae and McGregor, 1996). These surveys would establish the salient socio-economic characteristics of the communities as these affect reforestation efforts.

A successful outcome of the FORRU programme will be the adoption of good, appropriate nursery practices and the cultivation of a wider range of native tree species, as a result of guidelines which can be used by organizations in northern Thailand in their reforestation programmes.

Further Reading

F. HALLE, P. A. OLDEMANN and P. B. TOMLINSON (1978) *Tropical Trees and Forests*. Springer-Verlag, New York.
R. M. KOOYMAN (1996) *Growing Rainforest: Rainforest Restoration and Regeneration*. Australia: Greening Australia (NSW) Inc.
E. SHANKS and J. CARTER (1994) *The Organization of Small-Scale Tree Nurseries*. Rural Development Forestry Study Guide 1. Overseas Development Institute, Oxford.
W. R. SORDEN, M. E. GILPIN and J. D. ABER (1987) *Restoration Ecology: a Synthetic Approach to Ecological Research*. Cambridge: Cambridge University Press.

References

S. BHUMIBAMON (1986) *The Environmental and Socio-economic Aspects of Tropical Deforestation: a Case Study of Thailand*, Department of Silviculture, Faculty of Forestry, Kasetsart University. Thailand.
D. BLAKESLEY, N. PASK, G. G. HENSHAW and M. FAY (1996) 'Conservation of forest genetic resources; in vitro strategies and cryopreservation', *Plant Growth Regulation*, 20, pp. 11–16.
B. BOONTAWEE, C. PLENGKLAI and A. KAO-SA-ED (1995) 'Monitoring and measuring forest biodiversity in Thailand', in Boyle, T. J. B. and Boontawee, B. (eds), *Proceedings of a Symposium on Measuring and Monitoring Biodiversity in Tropical and Temperate Forests*, Chiang Mai Thailand, CIFOR. pp. 113–26.
S. ELLIOTT, V. ANUSARNSUNTHORN, N. GARWOOD and D.

BLAKESLEY (1995) 'Research needs for restoring the forests of Thailand'. *Nat. Hist. Bull. Siam. Soc.* 43, pp. 179–84.

FAO, 1997. *State of the World's Forests 1997.*

P. GAVINLERTVATANA (1992) 'Micropropagation of bamboo'. *Tigerpaper* 29, pp. 1–9.

K. HARDWICK, J. HEALEY, S. ELLIOTT, N. C. GARWOOD and V. ANUSARNSUNTHORN (1997) 'Understanding and assisting natural regeneration processes in degraded seasonal evergreen forests in northern Thailand', *Forest Ecology and Management* 99, pp. 203–214.

P. HIRSCH (1990) 'Forests, forest reserve, and forest land in Thailand', *The Geographic Journal* 156, pp. 166–74.

M. KANTARLI (1993) *Vegetative Propagation of Dipterocarps by Cuttings in ASEAN Region*, Review Paper No. 1. ASEAN-Canada Forest Tree Seed Centre, Muak Lek, Thailand.

S. KIJKAR (1991a) *Handbook: Coconut Husk as a Potting Medium*, ASEAN–Canada Forest Tree Seed Centre, Muak Lek, Thailand.

S. KIJKAR (1991b) *Handbook: Producing Rooted Cuttings of* Eucalyptus camaldulensis, ASEAN–Canada Forest Tree Seed Centre, Muak Lek, Thailand.

S. KOPACHON, K. SURIYA, K. HARDWICK, G. PAKAAD, J. F. MAXWELL, V. ANUSARNSUNTHORN, N. GARWOOD, D. BLAKESLEY and S. ELLIOTT (1996) 'Forest restoration research in Northern Thailand. 1. Fruits, seeds and seedlings of *Hovenia dulcis* Thunb', *Nat. Hist. Bull. Siam. Soc.* 44, pp. 41–52.

B. KRISHNAPILLAY (1992) *Development of Seed Testing Standards for* Dipterocarpus alatus, D. intricatus *and* Hopea oderata. Technical Pub. No. 15, ASEAN–Canada Forest Tree Seed Centre, Muak Lek, Thailand.

D. LAMB, J. PARROTTA, R. KEENAN and N. TUCKER (1997) 'Rejoining habitat fragments: restoring degraded forest lands'. In Laurance, W. F. and R. O. Bierregaard, *Tropical Forest Remnants: Ecology, Management and Conservation of Fragmented Communities*, pp. 366–385.

P. LEUNGARAMSRI and N. RAJESH (1992) *The Future of People and Forests in Thailand After the Logging Ban*, Project for Ecological Recovery, Bangkok, Thailand.

A. MASAE and J. A. McGREGOR (1996) *Sustainability Under Conditions of Rapid Development: a Freshwater Fishery in Southern Thailand*, CDS Occasional Paper, 01/96.

J. F. MAXWELL, S. ELLIOTT and V. ANUSARNSUNTHORN (1995) 'The vegetation of Doi Khuntan National Park, Lamphun-Lampang Provinces, Thailand', *Nat. Hist. Bull. Siam Soc.* 43, pp. 185–205.

J. F. MAXWELL, S. ELLIOTT and V. ANUSARNSUNTHORN (1997) 'The vegetation of Jae Sawn National Park-Lampang Provinces, Thailand', *Nat. Hist. Bull. Siam Soc.* 45, pp. 71–97.

T. SUKPANICH (1998) Stumped? Bangkok Post 14/6/98 p. 6.

World Conservation Monitoring Centre (1996) *Guide to Information Management in the Context of the Convention on Biological Diversity*, Nairobi: United Nations Environment Programme.

13 Sustainability, Access and Equity: Tropical Floodplain Fisheries in Three Asian Countries

Christopher J. Heady, J. Allister McGregor and Adrian Winnett

Summary

Tropical inland fisheries are an important source of food, employment, and income for many poor people in rural Asia. Compared to marine fisheries there has been little systematic research to investigate their biology or socio-economic organisation; rather generalized assertions are made which assume that their sustainability is threatened, especially by weak control of access which leads to overfishing. Policy conclusions are then based on these assertions. The present chapter reports a large scale ODA-financed project which investigated three floodplain fisheries, in Bangladesh, Thailand, and Indonesia from 1992 to 1994. The project was interdisciplinary, drawing on the expertise of biologists, economists, and other social scientists, and used a variety of quantitative and qualitative techniques to analyse the fisheries. It was particularly concerned to view the fisheries not just in terms of their biology or their social economy, but to show how these interacted: here we focus on the results of some computer simulations of various management scenarios. These show that there is little evidence of generalized overfishing. Thus the standard policy responses may yield little in terms of enhancing sustainability, but they potentially have quite serious micro-level effects on the incomes of particular households. Contrary to some assertions, there are sophisticated, traditional controls on access to these fisheries, but these are dependent, in part at least, on acceptance of their outcomes in income-distributional terms. Policy-makers therefore need to be alert to the implications of their actions for income distribution to avoid undermining such controls on access. More generally, threats to sustainability in a biological sense seem more likely to come from environmental shocks rather than from the internal dynamics of effort adjustment in the fishery, and in a socio-economic sense from changes in the wider economy, society, and polity which place pressures on the traditional management of the fisheries. Both these are, of course, manifestations of broader processes of development. Ill-judged policy responses, however well-intentioned, may contribute to these threats.

I INTRODUCTION: THE PROJECT

This chapter sets out some of the findings of an interdisciplinary research project (Heady *et al.*, 1995) which studied inland capture fisheries in Bangladesh, Indonesia and Thailand. The main objective of this research was to explore the extent to which the sustainability of small-scale inland fisheries was under threat in the three countries. This involved the project in establishing the biological status of the three fisheries; exploring both the formal and informal management practices at each of the water-bodies; and considering the interplay between the pressures created by the quite different development experiences of the three countries and the exploitation of this valuable renewable natural resource. As is well recognized, in South and Southeast Asia freshwater fish are an important source of relatively cheap protein; such fisheries also provide income and employment for significant numbers of rural people, underpinning the way of life of many small (and some not-so-small) communities.

In Bangladesh, the poorest of the countries, the study site was Hail Haor, a large flooded depression in the Northeast province of Sylhet; in Indonesia, the site was the Lempuing river, and its associated backwaters, in South Sumatra; and in Thailand, the site was Thale Noi lake and its associated swamp, Phru Khuan Khreng (the 'Phru'), in the South of the country. All of these are freshwater systems, with extensive areas of permanent water, but with large seasonal variations in water levels and temporarily inundated areas. All three are what can be termed artisanal peasant fisheries: each is exploited on a predominantly household basis and with fishing gears which can usually be operated by a single fisherman. Indonesia is a minor exception to this in that a feature of the Lempuing River fishery is a number of large-scale, fixed gears which are operated for a season at a time by groups of fishermen. While the fisheries are still quite traditional in their appearance—with, at most, mechanisation restricted to simple outboard motors—this should not suggest a 'subsistence' type of activity. A striking feature of the three sites is that they are well integrated into complex and modern trading networks and fish caught in each may be traded commercially, often over quite large distances and even internationally.

Beyond this, there is considerable diversity in both the natural and social environments. In particular, we should note, for later reference, that in Thailand there is no formal control of access to the fishery through a system of private property rights; nonetheless, within a loosely enforced legislative framework there are community norms and customs regulating the fishery, of the sort familiar from other studies of common property resources. In Sumatra and Bangladesh there are short-term leases giving rights to fish over specified areas of water, and these are periodically sold at auction; but these private property rights are very incomplete, and are themselves embedded in other cultural and social norms and customs. Thus the distinction between the access regime in Thailand, on the one hand, and Sumatra and Bangladesh, on the other, is not as clear as it initially appears to be.

The project observed a complete annual cycle of fishing activity at each of the fisheries during the period from late 1992 to early 1994. The observation had biological and socio-economic components. The former built up a detailed profile of the exploited fish stocks: their composition by species and age, and their mortality rates from catches by different sorts of fishing gear. The latter involved keeping detailed records of the fishing and other activities of a sample of households, recording patterns of income and employment week by week; this information was collected and analysed by both economists and anthropologists. Together with background information on the natural environment, the formal and informal regimes under which the fisheries are managed, and the markets in which fishing households operate, the project provided a detailed picture of the way in which fisheries of this type operate. Indeed, it is probably the most detailed available.

Much of the work of project consisted of assembling this information into a coherent analysis; in part, this involved the construction of a formal, computable model, but there are many important aspects of these fisheries which cannot be captured in models of this sort. In several respects, we have been led to depart from much of the received wisdom (outlined in the next section of this chapter) on the nature of the problems which confront those who seek to analyse and manage inland tropical fisheries. In the present chapter, we focus on this questioning aspect of our work, and, especially, draw on our formal model to illustrate the argument.

II THE RECEIVED WISDOM: 'OVERFISHING' AND THE 'TRAGEDY OF THE COMMONS'

Floodplain fisheries are exceptionally diverse and complex in biological terms, but, in contrast with marine fisheries and with aquaculture, inland capture fisheries have been little studied, and many of the presumptions about them—routinely drawn from models of large-scale, commercial marine fisheries (and these often for temperate waters)—seem to be misleading. In particular, the concepts of 'overfishing' and the 'tragedy of the commons' (Hardin, 1968, but earlier and more coherently applied to fisheries by Gordon, 1954) have long been central to what has become the received wisdom on the problems of fisheries. Examination of fisheries policies in the three countries and discussions with policy-makers indicated that these concepts have influenced official management strategies towards small-scale inland fisheries in the countries of South and Southeast Asia (as elsewhere in the developing world), even though there is no serious body of empirical evidence to confirm or disconfirm the applicability of the concepts. These concepts were also influential in the initial formulations of the objectives of this research project. However, as the project progressed, it became apparent that in many ways this guidance was somewhat misleading. The real issues which confront the types of fishery studied are in some respects less serious than the received wisdom might suggest, but in others perhaps more serious—or, at least, more complex and subtle.

There are, essentially, four interrelated features of these fisheries which underpin this assertion:

(1) the multi-species and multi-gear character of floodplain fisheries of this kind;
(2) the strong seasonality in the pattern of fishing activity;
(3) the fact that the fisherfolk do not constitute a homogeneous group;
(4) the nature of the links which exist between various sub-groups

in fishing communities and the wider economy, polity and society.

Put thus baldly, these statements may appear trite, but unless they are taken very seriously indeed, as basic components in our understanding of the fisheries, we are likely to be led astray in both analysis and policy. In contrast, the received wisdom often treats such features as secondary complications. (For a similar general argument, see Wilson, 1982.)

These four characteristics underpin the dynamics of income distribution within the types of fishing communities studied. As we explain shortly, income-distributional relationships are central to understanding the problems confronted by these communities, and especially to the success or failure of policies designed to mitigate their problems; but this distributional question is an issue little considered in most discussions of fisheries policy.

Such discussions, drawing on the received wisdom, are very much orientated toward understanding the problem of 'overfishing'. Overfishing may be understood in two senses, as 'too low' levels of fish stocks or 'too high' levels of effort employed in fishing. These are, of course, related: we would generally expect that more effort is required to achieve a given catch the lower the fish stock. Sometimes overfishing is taken as evidenced simply by declining fish stocks and rising fishing effort over time, possibly summarised by declining catches *per* unit of effort. Even if the observation is empirically correct, this is too casual an inference, though it has helped to shape perceptions of many fisheries, including those in the countries studied. Catch *per* unit of effort may decline for a variety of reasons, including environmental changes, and it may, in any case have been 'too high' initially in economic terms.

It is important to realize that, strictly speaking, overfishing is always defined relative to some model of how the fishery operates. Overfishing in most standard models arises either because access to the fishery is not properly controlled or because future net yields from the fishery are discounted at an inappropriately high rate. The former entails excessive fishing effort and therefore inadequate fish stocks, the latter inadequate stocks and therefore excessive effort, both relative to what is economically efficient. This rather more sophisticated understanding of the nature of overfishing has also shaped policy-makers' and managers' perceptions of many fisheries, again including those in the countries studied. It is often assumed that, in the absence of well-defined property rights to control access to the fishery, effort levels will be excessive, and that, in the absence of well-functioning financial markets to set appropriate interest rates at which to discount future yields, fish stocks will be too low. The former is usually taken to be the major cause of the 'tragedy of the commons', though this notoriously amorphous concept often includes elements of the latter too. (Poorly controlled access is often assumed to lead to excessive subjective discount rates among those who exploit the fishery, since individual fishing units have no apparent stake in the future of the fishery.) Ill-defined property rights and ill-functioning financial markets are often assumed to be endemic in developing countries.

The standard policy conclusion that is drawn from all this is the need for more clearly defined and monitored property rights, and especially taking the financial market dimension into account, generally more 'marketization' of the fishery. (Or, alternatively, for various types of regulation to mimic these effects; though popular with fisheries' officers, these are out of fashion with the higher echelons of policy-makers, especially in international agencies.)

Though we have presented this familiar argument in its simplest form, it is clearly dependent on numerous assumptions; much of our work of the project has led us to question at least some of these assumptions, especially those relating to access and property rights, but more tentatively also those relating to financial markets. In particular, we note again the importance of norms and customs in regulating both the apparent open access in Thailand and the apparent private property rights in Bangladesh and Sumatra. Thus it may be misleading to work with assumptions which draw too clear a contrast between open access and well-defined property rights; there is partial recognition of this in the more subtle use of the term 'the commons' in much recent discussion. These might be described as socio-economic assumptions.

But there are also what might be described as bioeconomic assumptions behind the received wisdom—essentially that aggregate catch is sensitive to variations in fishing effort over some appropriate range, as assumed by most standard bioeconomic models of fisheries, including that outlined in the next section of this chapter. However, the empirical validity of this assumption needs to be established for each case.

It is important to understand that the empirical relevance of these two groups of assumptions can be separately questioned, but that a more convincing picture of how the fisheries operate can be presented by bringing the two sets of questions together. The reason for this is that there is a close link between access to the fishery and the distribution of income from the fishery, both among those who directly work the fishery and between them and other claimants on the income of the fishery (including the agencies and agents of the state). This is always the case, though the socio-economic complexity of the fisheries studied makes this access-distributional nexus especially complicated and constraining. However, the received wisdom sees the solution to overfishing as an efficiency gain to be traded-off against the disruption of the pattern of income distribution which would result from a tighter restriction on access. (And a disruption which, if thought undesirable, could be handled by compensatory policies: for example, by adjusting tax burdens to redistribute income.) But if the aggregate catch is relatively insensitive to variations in effort over wide ranges, there may be little efficiency gain to be found, and all that is left is the distributional disruption. And, especially in the context of developing countries, it is probably sanguine to believe that compensatory income redistribution is feasible. Limitations of information and administration forcibly suggest that it is not. This, put strongly, seems to be the case for the fisheries studied. (And not only for these, as a moment's reflection on the problems of fishing in British coastal waters will confirm.) Much of the rest of this chapter is concerned with substantiating these arguments.

Before moving on, we should emphasise that we are not saying that there is never likely to be a problem of long-term sustainability in the type of fisheries studied. Rather, if there is a problem it is likely to come from pressures external to the fishery, rather than from, as in the received wisdom, the inadequate management of fishing effort within the community. These may be pressures which adversely shift the whole catch–effort relationship or disrupt the legitimacy of the pattern of distributional claims on the fishery. The first may affect sustainability directly, but, by undermining the customary access-distribution nexus, both may set in motion moves to altogether different modes of exploitation which may not be sustainable. We return to this in our conclusion.

III   BIOECONOMICS: MODELLING CATCH AND EFFORT IN THE FLOODPLAIN FISHERY

We identified as one of the leading features of these fisheries their multi-species and multi-gear character. In all three fisheries, there are large numbers (30–40) of species caught by equally extensive ranges of active and passive fishing gears, which vary in their selectivity among species, according to factors such as location, time of use, and mesh size. Most of the fish species have high rates of reproduction, growth, and mortality, and depleted species are quickly replaced by others. All of this reflects adaptation to the high fertility and variability of the floodplain environment.

Because of the high quality of the data available, the project has been able to capture relevant aspects of much of this complexity by adapting and extending a model, known as BEAM4, originally developed by the FAO. The most important result of this model, for the fisheries studied, is to show the relative insensitivity of catch to a wide range of general and selective variations in fishing effort. This formally confirms our more intuitive understanding and is in conformity with other evidence from floodplain fisheries (for example, Welcomme, 1985).

There is an important preliminary point which must be emphasised. Models of this type are based on predicting yield *per* recruit to the fish stock, which is assumed to be proportional to total yield. If fish stocks are so depressed that recruitment fails, then the fishery fails. This has not so far been observed for the study sites, and the assumption must be that stocks are adequate for recruitment. But very little is known about what is 'adequate' in fisheries of this type. This is a subject which requires further investigation, but we suspect that this is just the sort of problem that could be generated by the external environmental pressures mentioned above, such as declining water depths or quality, which could have simultaneous effects across many species. (Or, as seen in some marine fisheries, by fishing methods which themselves destroy the environmental support, such as a coral reef, of the fishery. If industrialized fishing spreads to inland waters this could become a problem there too, as is apparent from certain kinds of aquaculture.)

BEAM4 is essentially a multi-gear and multi-species version of the cohort-based dynamic pool model widely used in fish biology. The simple versions of such models assume that each species of fish grows asymptotically towards its maximum length (and weight), and has an instantaneous mortality rate due to natural mortality and to fishing mortality. Much of the biological sampling programme was aimed at establishing these parameters. (Given the large number of species, these are aggregated into a small number of 'guilds' appropriately defined by conditions at each site.) At low levels of fishing effort, natural mortality is relatively high, and at high levels of effort, fishing mortality is relatively high, thus catch (in terms of weight) and effort generally have the standard type of relationship outlined at the end of Section II. As already suggested, this simple relationship effectively disappears in the more complex case considered here.

The BEAM4 model is initially fitted using best estimates of growth, mortality, and—here is the crucial emendation from the simple dynamic pool model—gear selectivity among species. It is then adjusted to give the closest possible reference simulation of observed patterns of catches. Management scenario simulations can then be carried-out, estimating long-run outcomes on the assumption that recruitment is unaffected. These scenarios correspond to the sorts of recommendations often made by fisheries officers, but, especially in the case of (1), mimic the sort of adjustments to fishing effort that might be made by the rent-maximizing exclusive owners beloved of neo-classical economics.

The scenarios considered were for:

(1)　changes of effort, in terms of labour time, across all gears;
(2)　closures of certain gears or types of gear;
(3)　closed seasons for some or all gears;
(4)　limits on mesh size (which can also approximate limits on size of fish to be caught).

The advantage of this type of modelling is that it can potentially account for complex patterns of interaction: at the simplest, closure of one type of gear may increase catches to other types. Since in this and the following section, we will be quoting extensively from the results of computer simulations, we should say a little about how much weight we place on such predications, and, more broadly, about how we approach the question of formal modelling. First, the key results are very clear-cut and robust; we are not in the realm of dubious inferences. Secondly, we regard the predictions primarily as a way of drawing illustrative interpretations from our extensive quantitative fieldwork; throughout we are aware that they have to be placed in the context of a more qualitative understanding of the ways in which these fisheries operate.

With this background, we present some summary results in Tables 13.1–13.3. These give aggregate weight and value of catches. It should be noted, though, that this conceals some variability in the behaviour of catches from individual guilds. These underlying results show that, at each study site, some species are overexploited while others are

Table 13.1 Changes in yield from alternative management scenarios for the Hail Haor fishery

Management scenario	New catch as % of reference catch	
	In weight terms	*In value terms*
Total Ban on		
Seine nets	99	100
Dol brushpiles	101	102
Felun pushnets	96	97
Gill nets	96	98
Dori traps	89	90
Faron traps	99	99
Bosni traps	99	99
Small hooks	93	94
Large hooks	100	100
Dry fishing	100	100
Effort Changes		
All gears × 25%	61	68
All gears × 50%	83	87
All gears × 75%	94	95
All gears × 125%	104	103
All gears × 150%	106	104
Closed Seasons		
June–Sept.	93	95
Oct.–Dec.	94	95
June–Dec.	78	82
June–Dec. (carps only)	102	104
Mesh Size Increase		
To size of gill nets	86	87
To 1.5 × gill nets	57	63

underexploited. Thus a reduction in fishing effort would increase both the physical stock and the yield of some species. Along with evidence about the maturity of fish caught, this is often used as a more subtle argument to show the existence of overfishing in some cases.

However, from the point of view of providing an income source to the fisherfolk, what is important is the total value of the catch. Our socio-economic data can be combined with the biological data to generate results in terms of value of catch, estimated at current market prices. Tables 13.1–13.3 show that the scenarios generally have only a small effect on either the total weight of fish caught or the total catch value. For example, at Hail Haor, the only restrictive scenarios that

Table 13.2 Changes in yield from alternative management scenarios for
the Lempuing fishery

Management scenario	New catch as % of reference catch	
	In weight terms	*In value terms*
Total ban on		
Floodplain barriers	99	100
River barriers	102	101
Ngesar seines	100	100
Bumbun FAD/seine	100	100
Gill nets	99	99
Rattan/wire traps	96	98
Bamboo traps	98	91
Lapun traps	100	100
Seruo traps	102	102
Hooks	102	102
All group gears	90	93
All individual gears	90	85
Effort changes		
All gears × 50%	96	94
All gears × 75%	100	99
All gears × 125%	99	99
All gears × 150%	97	97
Group gears × 150%	98	97
Individual gears × 150%	99	99
Closed seasons		
Jan.–Feb.	100	98
Jan.–Mar.	100	95
Jan.–Apr.	101	97
Jan.–May	102	98
Jan.–June	103	99
Jan.–July	100	96
Jan.–Aug.	95	93
Jan.–Sept.	88	88
Jan.–Oct.	68	69
Jan.–Nov.	21	22
Mesh Size Increase		
To bamboo traps	102	104
To 1.5 × bamboo traps	86	92

increase the catch are one gear ban and one (selective) closed season,
and these by very small amounts. These results are not consistent with
the view that the fishery is generally overexploited Given the usual
perceptions of resource use in Bangladesh this is a surprising result.

Table 13.3 Changes in yield from alternative management scenarios for the Thale Noi/Phru fishery

Management scenarios	New catch as % of reference catch	
	In weight terms	*In value terms*
Total ban on Thale Noi fisherfolk		
Seine nets	102	104
Gill nets	80	90
Fish traps	101	102
Hooks	102	104
Trap ponds	100	101
Total ban on Phru fisherfolk		
Seine nets	100	100
Gill nets	100	101
Fish traps	100	100
Hooks	101	102
Trap ponds	100	101
Total ban on		
All seine nets	102	104
All trap ponds	101	102
All fish traps	101	102
Effort changes		
All gears × 50%	96	103
All gears × 75%	100	103
All gears × 125%	98	96
All gears × 150%	96	92
Closed seasons		
Oct–Jan., except traps	101	104
Oct.–Jan., all gears	101	106
Jan.–Feb.	99	101
Jan.–Mar.	98	101
Jan.–Apr.	97	101
Jan.–May	95	100
Jan.–June	92	98
Jan.–July	89	96
Jan.–Aug.	85	93
Jan.-Sept.	76	85
Jan.–Oct.	58	66
Jan.–Nov.	33	41
Mesh size increase		
To gill nets	105	108
To 1.25 × gill nets	110	119
To 1.5 × gill nets	104	121

Indeed, small increases in total yield are generated by increases in fishing effort. Similarly, the Lempuing fishery in Indonesia also shows no substantial yield increases from more restrictive management.

The picture is slightly different for the Thai fishery, where more restrictive policies do seem to increase total yield. It is tempting to conclude that this is evidence of overfishing, and that it is due to the fact that there is apparent open access. However, the accuracy of the estimates is not sufficient to allow us to place reliance on numbers that are generally only slightly larger than 100%. We can only safely conclude that there is no evidence of serious general overfishing in any of our three study sites, although there is overfishing of some species (and of some maturities).

IV SOCIO-ECONOMICS: ACCESS AND INCOME DISTRIBUTION IN THE FISHING COMMUNITY

It is clear from our work, as illustrated in Tables 13.1 to 3, that very large changes in labour time, beyond anything that could be regarded as realistic, are necessary to produce more than small changes in the value of catch. Suppose we were to think of the fishery as under the control of a rent-receiving owner. The main effect of the owner restricting access would be to restrict employment and the labour incomes of fisherfolk at the expense of rent. There would be very little effect on the total value of catch, but much distributional disruption.

This case is very important in understanding the situation in Bangladesh and Indonesia, since, as mentioned, through a system of auctions of fishing rights, there is indeed a partial system of private ownership of the fishery by leaseholders. There is a temptation to view this as an inadequate property right, awaiting enhancement as part of a marketization programme. As we have explained in more detail elsewhere (Winnett and McGregor, 1994), this is a dangerous misinterpretation. The auctions of fishing rights are only a part of wider community involvement in the management of access to the fishery, and, as always, this legitimates and is legitimated by its income distributional characteristics.

In effect, we suggest that the observed levels of labour usage may be socially desirable, and, to some extent, this is embodied in the community's norms and customs to which leaseholders respond. The fact that employment creation is seen as valuable suggests that the wage is higher than the social opportunity cost of employment. This is certainly so in Bangladesh, but we should note that the social functioning of the leasing system is under pressure from the official and unofficial revenue demands of an over-extended state.

The extent of underemployment in Indonesia is not as high as in Bangladesh; however, there may well still be social benefits from employment creation. On the Lempuing, reduced employment in the fishery can be expected to lead to people leaving the community to find alternative employment. Some migration of this sort is an inevitable part of economic development, but excessive migration can lead to serious social problems in both the community that is losing people and the community that is receiving them. Employment in the fisheries helps to moderate the flow of migrants. However, increasing integration of markets for fish has encouraged long-distance traders to compete for leases in Sumatra; and this seems to be weakening the role of local leaseholders and, relatedly, the wider social function of the auction system. Comparable pressures are apparent in Thailand, but not, of course, manifested through a leasing system. This, once more, emphasises the destructive potential of changes external to the fishing community.

The distributional issue is more complex than the simple division between labour and rent income observable in the leasing systems of Sumatra and Bangladesh. In all three countries, different fishing gears are owned and operated by different social groups, so any consideration of changing the level or pattern of effort in the fishery should include an analysis of their income distributional effects. Such an analysis needs to include both short-run and long-run effects. In the short-run, restricting use of a particular gear will reduce the income of those who normally use that gear and increase the income of others who normally use gears whose catches have increased as a result. However, as time passes, there will be responses to this initial change.

In Indonesia and Bangladesh leaseholders might increase the charges ('tolls') for the use of the gears that have become more profitable, and reduce them for gears that have become less profitable. They might also seek to increase the use of gears that have become more profitable and reduce the use of gears that have become less profitable. The effects of these changes on employment and labour income will depend on the labour intensities of the gears with increasing use as compared to those with decreasing use. In Thailand, the responses will not be dictated by leaseholders but will come from

individual fisherfolk, who are constrained only by the capital requirements of purchasing gears and by the limited availability of suitable
fishing sites. There will be a general move from the gears that have
had their catches reduced to those that have had increases.

In general, the long-run adjustments will reduce the sizes of the
short-run income gains and losses to particular groups of fisherfolk,
replacing them with changes in employment. But it may take some
considerable time for the long-run adjustments to operate.

The short-run effects can be calculated by combining our estimates
of catch changes for specific gears with our socio-economic data on
fishing households. Our surveys report the amount of each household's income that is obtained from the use of each fishing gear. For
each scenario, we apply to each household's income from each gear
in each time period the proportionate change in catch for that gear
that is predicted by the model. By summing over all gears, for each
household, we can then estimate the change in income for each household. It is important to note that the scenarios alter the composition
of the catch taken by each gear. The effect on the household's income
must therefore be calculated by using the proportionate change in the
value of the catch, rather than the proportionate change in weight.
(We were not able to apply this method to the data from Indonesia
as the nature of group fishing on the Lempuing prevented us from
fully allocating fishing income amongst the households. In Thailand,
it was not possible to apply the method to the Phru fisherfolk, because
of incomplete data.)

In Bangladesh and for Thale Noi in Thailand the households that
derived a significant income from fishing were divided into three
groups (high income, middle income and poorest) according to our
survey estimates of their relative wealth (based on giving points for
various classes of possessions). The effects of each scenario were calculated for each household and the averages for each group are shown
in Table 13.4. In addition, the effects on the worse affected and best
affected households in the group are shown in parentheses. So, for
example, a seine net ban in Hail Haor, shows the poorest households
averaging a 1% fall in income, with the changes ranging from a fall of
36% to an increase of 8%.

These results show that the effects of the various scenarios on the
average income of these groups is only slightly larger than their effects
on the total catch value. In contrast—and this is the important point—
there are very often large distributional effects within each group. In
other words, it makes little sense to talk of the effect of changing pat-

Table 13.4 Distributional effects of alternative management scenarios for the Hail Haor and Thale Noi/Phru fisheries

Management scenario	New income as % of current income		
	High income	*Middle income*	*Poorest*
Hail Haor			
Ban on seine nets	107 (93,123)	99 (65,111)	99 (34,108)
Ban on brushpiles	97 (31,118)	102 (67,121)	107 (99,121)
Ban on gill nets	103 (32,171)	109 (29,145)	97 (40,132)
50% effort reduction	91 (75,141)	89 (71,125)	87 (75,104)
50% effort increase	101 (81,111)	102 (81,113)	103 (88,116)
June–Sept. closure	99 (67,148)	93 (51,133)	100 (41,215)
June–Dec. closure	86 (49,181)	74 (17,121)	61 (15,147)
Thale Noi			
Ban on seine nets	100 (50,119)	104 (50,120)	91 (0,120)
Ban on fish traps	93 (0,123)	89 (0,120)	105 (27,123)
Ban on hooks	104 (65,125)	91 (30,130)	104 (0,129)
Ban on trap pond	102 (98,103)	100 (92,103)	102 (80,103)
25% effort reduction	102 (99,106)	103 (99,107)	101 (95,110)
Oct.–Jan. closure, Except traps	107 (100,137)	111 (100,137)	110 (87,149)
Oct.–Jan. closure, All gears	112 (55,163)	115 (84,155)	114 (73,173)
Mesh size increase to 1.25 × gill net	116 (100,137)	117 (100,137)	111 (87,149)

terns of fishing activity on 'poor fisherfolk' as a uniform group. One has to look at individuals. Thus concealed within fairly stable aggregates and averages are large potential shifts in income at the micro level: attempts to change levels or patterns of fishing activity could have serious distributional effects.

V CONCLUSION: COMMUNITIES AND SUSTAINABILITY

In section 2 of this chapter we gave a four point characterisation of the fisheries studied. We have now summarized some of the information on which this was based, and we have seen that it supports two major conclusions:

(1) Total catch values tend to be resilient against many types of effort variation.

(2)   Existing local management regimes, particularly those relating
to access, have a strong income-distributional dimension. Their
acceptability to the communities depends largely on how the
legitimacy of these distributional arrangements is perceived.

We recall that these are, strictly, independent arguments, but they
are complementary. They identify, very clearly, the limits of the trade-
off between equity and afficiency as the key question in understanding and appraising the organisation of these fisheries.

The fisheries literature has tended to focus on overfishing as the
problem to be addressed, and of low incomes in fishing as, in part, both
cause and effect of this. Such a view is based essentially on, so to speak,
the internal dynamics of the fishery. Our view is, rather, that the sustainability problems of the fishery are better understood as part of the
wider problems of the economy, society, and polity of which it forms
a part. Thus:

(1)   Threats to biological sustainability are most likely to arise from
environmental degradation, particularly if this leads to recruitment failures.
(2)   Pressures on the viability of the local management regime are
most likely to arise from external pressures on the system of
income entitlements within the fishery.

We focus on the second of these. Such pressures may arise because,
for example, of changing patterns of economic opportunities in the
wider socio-economy: both too few opportunities (Bangladesh) and
too many opportunities (Thailand, and to a much less noticeable
extent at present, Indonesia) can cause problems. In their different
ways, rural overcrowding and outward migration can be disruptive of
rural fishing communities in terms of shifts in the size and structure
of population which then feed through into patterns of income distribution. Especially in the case of Bangladesh, such disruption is
aggravated through an overstretched state increasing its revenue
demands on immobile and identifiable objects of taxation, such as
natural resources. Further, both rural overcrowding and outward
migration share some common causes with environmental degradation, when related to the wider contexts of extreme population pressure and rapid, unbalanced economic growth, respectively.

Thus any attempts to intervene in the management of the fishery, in the belief, say, that these interventions are necessary to maintain biological sustainability, have to be alert to the capacity of the local management regime to absorb distributional shocks and to whether the best solution lies within the fishery at all. A first question to be asked about the income entitlements in the fishery is how maintainable the weakest of them are likely to be if there are enhanced restrictions on access. Much of what is perceived as socially legitimate in the way the fishery is organised probably flows from its capacity to deal with such micro-distributional problems.

Consider perhaps the leading example, but one representative of the changes affecting much of East Asia. In Thailand, given the evolution of the wider socio-economy, for any kind of fishery to maintain itself in the long-run would probably require full-scale commercialization in order to justify the required capital investment. But this would no longer be recognisable as the present socially integrated community. Under this new regime, the internal dynamics of the fishery may then, indeed, threaten sustainability: there are levels of technical innovation which may undermine even the most resilient resource. (There are already signs of this in aquaculture.) In short, the sustainability of the resource cannot be separated from the sustainability of the community that works it. Our argument is that community sustainability is crucially dependent on perceived distributional legitimacy, and, in turn, distributional legitimacy is tied to resource sustainability through control of access.

VI TOWARDS AN ENVIRONMENT RESEARCH AGENDA

Fisheries research has been the province of specialised groups of biologists and economists. It is generally true to say that the elaboration of models on the biological side has not been matched by comparable elaboration on the economic side, and vice versa. However, our work has shown that there is an important interaction between complexity in the biological structure of the fish stock, with multiplicity of species and maturities, and complexity in the adaptations of effort available to fisherfolk, along dimensions which include selection of gear types and of location and time of fishing. Thus it is necessary to go beyond simple homogenous measures of fish stock and fishing effort, and this will require more extensive collaboration and understanding between biologists and economists, continuing a process which we began in this project.

Wider interdisciplinary collaboration is also called for. As the argument of this chapter makes clear, many of the problems of fisheries cannot be understood independently of the wider physical environment and of the society, economy, and polity of which they are part. Again, the project made a start here, with expertise drawn from anthropologists and policy analysts, alongside fish biologists and economists, but this range needs to be further extended to include natural scientists other than fish biologists, as well as other social scientists, and to be more fully integrated into our understanding of fisheries, and of other resources.

Acknowledgements

The project was financed by the UK Overseas Development Administration as part of their Natural Resources and Environment Division's initiative on the socio-economic dimensions of renewable natural resources. Biological sampling and BEAM4 modelling were undertaken by MRAG Ltd. Consultants, associated with Imperial College, London.

Further Reading

R. M. AUTY (1995) *Patterns of Development: Resources, Policy and Economic Growth*, London: Edward Arnold.
F. BERKES (ed.) (1989) *Common Property Resources: Ecology and Community Based Sustainable Development*, London: Belhaven.
J. M. BALAND and J. P. PLATTEAU (1996) *Halting Degradation of Natural Resources: Is There a Role for Rural Communities*, Oxford: Clarendon Press.
P. DASGUPTA (1993) *An Inquiry into Well-Being and Destitution*, Oxford: Clarendon Press.
J. M. HARTWICK and N. OLEWILER (1986) *The Economics of Natural Resource Use*, New York: HarperCollins.
R. B. NORGAARD (1994) *Development Betrayed*, London: Routledge.
E. OSTROM (1990) *Governing the Commons: the Evolution of Institutions for Collective Action*, Cambridge: Cambridge University Press.

References

H. S. GORDON (1954) 'The economic theory of common property resources', *Journal of Political Economy* 62, pp. 124–42.
G. HARDIN (1968) 'The tragedy of the commons', *Science* 162, pp. 1243–48.

C. J. HEADY, J. A. MCGREGOR and A. B. WINNETT (1995) *Poverty and Sustainability in the Management of Inland Capture Fisheries in South and Southeast Asia*. End of Project report for ODA Project Research Grant R4791, Centre for Development Studies, University of Bath.

R. L. WELCOMME (1985) *River Fisheries*. FAO Technical Paper 262, Rome.

J. A. WILSON (1982) 'The economical management of multispecies fisheries', *Land Economics* 58, pp. 417–34.

A. B. WINNETT and J. A. MCGREGOR (1994) *'The market meets the moral economy: the case of auctions in the management of small scale freshwater fishing in south and southeast Asia'*. Paper to SASE Conference, Paris.

Index